ALLE ZEIT WACH 1842

Springer Series in Synergetics Editor: Hermann Haken

Synergetics, an interdisciplinary field of research, is concerned with the cooperation of individual parts of a system that produces macroscopic spatial, temporal or functional structures. It deals with deterministic as well as stochastic processes.

1 **Synergetics** An Introduction 3rd Edition By H. Haken
2 **Synergetics** A Workshop Editor: H. Haken
3 **Synergetics** Far from Equilibrium Editors: A. Pacault, C. Vidal
4 **Structural Stability in Physics** Editors: W. Güttinger, H. Eikemeier
5 **Pattern Formation by Dynamic Systems and Pattern Recognition** Editor: H. Haken
6 **Dynamics of Synergetic Systems** Editor: H. Haken
7 **Problems of Biological Physics** By L. A. Blumenfeld
8 **Stochastic Nonlinear Systems** in Physics, Chemistry, and Biology Editors: L. Arnold, R. Lefever
9 **Numerical Methods in the Study of Critical Phenomena** Editors: J. Della Dora, J. Demongeot, B. Lacolle
10 **The Kinetic Theory of Electromagnetic Processes** By Yu. L. Klimontovich
11 **Chaos and Order in Nature** Editor: H. Haken
12 **Nonlinear Phenomena in Chemical Dynamics** Editors: C. Vidal, A. Pacault
13 **Handbook of Stochastic Methods** for Physics, Chemistry, and the Natural Sciences 2nd Edition By C. W. Gardiner
14 **Concepts and Models of a Quantitative Sociology** The Dynamics of Interacting Populations By W. Weidlich, G. Haag
15 **Noise-Induced Transitions** Theory and Applications in Physics, Chemistry, and Biology By W. Horsthemke, R. Lefever
16 **Physics of Bioenergetic Processes** By L. A. Blumenfeld
17 **Evolution of Order and Chaos** in Physics, Chemistry, and Biology Editor H. Haken
18 **The Fokker-Planck Equation** By H. Risken
19 **Chemical Oscillations, Waves, and Turbulence** By Y. Kuramoto
20 **Advanced Synergetics** 2nd Edition By H. Haken
21 **Stochastic Phenomena and Chaotic Behaviour in Complex Systems** Editor: P. Schuster
22 **Synergetics – From Microscopic to Macroscopic Order** Editor: E. Frehland
23 **Synergetics of the Brain** Editors: E. Başar, H. Flohr, H. Haken, A. J. Mandell
24 **Chaos and Statistical Methods** Editor: Y. Kuramoto
25 **Dynamics of Hierarchical Systems** An Evolutionary Approach By J. S. Nicolis
26 **Self-Organization and Management of Social Systems** Editors: H. Ulrich, G. J. B. Probst
27 **Non-Equilibrium Dynamics in Chemical Systems** Editors: C. Vidal, A. Pacault
28 **Self-Organization** Autowaves and Structures Far from Equilibrium Editor: V. I. Krinsky
29 **Temporal Order** Editors: L. Rensing, N. I. Jaeger
30 **Dynamical Problems in Soliton Systems** Editor: S. Takeno
31 **Complex Systems – Operational Approaches** in Neurobiology, Physics, and Computers Editor: H. Haken
32 **Dimensions and Entropies in Chaotic Systems** Quantification of Complex Behavior Editor: G. Mayer-Kress
33 **Selforganization by Nonlinear Irreversible Processes** Editors: W. Ebeling, H. Ulbricht
34 **Instabilities and Chaos in Quantum Optics** Editors: F. T. Arecchi, R. G. Harrison
35 **Nonequilibrium Phase Transitions in Semiconductors** Self-Organization Induced by Generation and Recombination Processes By E. Schöll
36 **Temporal Disorder in Human Oscillatory Systems** Editors: L. Rensing, U. an der Heiden, M. C. Mackey
37 **The Physics of Structure Formation** Theory and Simulation Editors: W. Güttinger and G. Dangelmayr
38 **Computational Systems – Natural and Artificial** Editor: H. Haken

Computational Systems – Natural and Artificial

Proceedings of the International Symposium
on Synergetics at Schloß Elmau, Bavaria
May 4–9, 1987

Editor: H. Haken

With 115 Figures

Springer-Verlag Berlin Heidelberg New York
London Paris Tokyo

Professor Dr. Dr. h.c. mult. Hermann Haken

Institut für Theoretische Physik, Universität Stuttgart, Pfaffenwaldring 57/IV,
D-7000 Stuttgart 80, Fed. Rep. of Germany and
Center for Complex Systems, Florida Atlantic University,
Boca Raton, FL 33431, USA

ISBN-13:978-3-642-73091-7 e-ISBN-13:978-3-642-73089-4
DOI: 10.1007/978-3-642-73089-4

Library of Congress Cataloging-in-Publication Data. International Symposium on Synergetics (1987 : Schloss Elmau, Bavaria) Computational systems–natural and artificial : proceedings of the International Symposium on Synergetics at Schloss Elmau, Bavaria, May 4–9, 1987 / editor, H. Haken. (Springer series in synergetics ; v. 38) Includes index. 1. System theory–Congresses. 2. Electronic data processing–Congresses. 3. Neural circuitry–Congresses. I. Haken, H. II. Title. III. Series. Q295.I586 1987 003–dc 19 87-28854

Softcover reprint of the hardcover 1st edition 1987

2153/3150-543210

Preface

This book contains the invited papers presented at an international symposium held at Schloss Elmau, Bavaria (FRG), May 4–9, 1987. Leading experts from neurobiology, medicine, physics, and the computer sciences joined together to present and discuss their most recent results. A particular example of the natural computational systems discussed is the visual system of man and animals. A bridge between neural networks and physical systems is provided by spin glass models of neural networks, which were also treated. Concrete realizations of new kinds of devices in microelectronics were among the further topics, as were general problems on the calculation of chaotic orbits. In this way these proceedings present a number of quite recent approaches to problems which are of great current interest in fields concerned with computational systems.

Bringing together scientists from neurobiology, physics, and the computer sciences has been one of the main aims of the synergetics enterprise, and in particular of its international symposia, from the very beginning. For example, its first meeting held in 1972 at Schloss Elmau included, among others, papers by R. Landauer and J.W.F. Woo on cooperative phenomena in data processing, by W. Reichardt on mechanisms of pattern recognition by the visual system of insects, by B. Julesz on stereoscopic depth perception, and by H.R. Wilson on cooperative phenomena in a homogeneous cortical tissue model. Whole meetings and the corresponding proceedings were devoted to these problems, e.g. "Pattern Formation and Pattern Recognition" (Springer Ser. in Synergetics, Vol. 5), where papers by K.S. Fu on syntactic pattern recognition and by T. Kohonen on associative memory are included. The proceedings on "Synergetics of the Brain" (Springer Ser. in Synergetics, Vol. 23), edited by Basar, Flohr, Haken and Mandell, contains papers by von der Malsburg, Bienenstock, Kohonen, Grossberg, Carpenter, Huberman and others on neural nets. The proceedings on "Complex Systems, Operational Approaches in Neurobiology, Physics and Computers" (Springer Ser. in Synergetics, Vol. 31) covers similar topics.

In a number of recent publications on the rapidly developing fields of neural computers and all-optical computers, the opinion was expressed that it is becoming more and more necessary to bring together neurobiologists, physicists, computer scientists and scientists from other branches of science. I am delighted that this type of meeting, which I initiated 15 years ago, has led to this kind of cross-fertilization and is now bearing fruit. I do hope that these proceedings in their turn will help to deepen our understanding of the complex processes of pattern recognition in natural systems and to pave the way for the construction of entirely new types of computers.

The organisation of these types of meetings in the frame of synergetics would not have been possible without the continued and generous support of the Volkswagenwerk Foundation. Their efforts are particularly remarkable because we are living in a time where, in general, support seems to be given by most agencies only to projects which are on highly specialized topics. Last but not least I wish to express my cordial thanks to Mrs. Ursula Funke who organized this meeting, as all the previous ones, perfectly.

Stuttgart, June 1987 *H. Haken*

Contents

Part I

Introduction

Synergetic Computers for Pattern Recognition and Associative Memory

H. Haken

Institut für Theoretische Physik und Synergetik, Universität Stuttgart, Pfaffenwaldring 57, D-7000 Stuttgart 80, Fed. Rep. of Germany

1. What these Proceedings are About

The purpose of my contribution is twofold: In this part I wish to present a brief preview on the contents of this volume and how its individual contributions are connected with each other. Then in the main part I wish to present my own model on the synergetic computer which allows for pattern recognition and associative memory. As I mentioned in the preface, it has been the goal of synergetics from the very beginning to bring scientists from neurobiology, physics, chemistry, computers, and other fields together and to provide a forum for interdisciplinary discussions. One of the main subjects of synergetics has been the relationship between pattern formation and pattern recognition. This tradition is followed up in the present proceedings where the first chapter deals with natural computational systems. Here we shall deal with the visual system (Koenderingk et al.), and (Baumgaertner et al.). (Here and in the following the quoted names refer to those authors who presented the paper.) It is wellknown from the cognitive sciences that we may complement an incomplete image, for instance by inserting automatically contour lines which are not present at all, but which are for instance indicated only by pieces of their ends. In the experiments by Baumgaertner et al. it is shown that under the above-mentioned circumstances quite evidently neurons start to fire which have been quiescent otherwise. Thus by means of the property of the total network of the visual cortex, there is a material substrate for this complementation effect. I do not doubt that these findings are of great importance for the construction of neural computers. An important problem in this context is figure / ground discrimination, on which a model is developed by Reitboeck et al. An abstract mathematical approach employing a variational approach is presented by Ingarden.

Macroscopic pattern formation in neural systems seems to occur in the E.E.G. patterns and two contributions (Lehmann et al. and Fuchs et al.) are devoted to this kind of phenomena. While the biological significance of E.E.G. pattern seems not yet clear at the present moment, we can analyse these formations as that of macroscopic complex systems in analogy to many other processes going on in other synergetic systems when selforganisation starts.

The line of thought how to analyse complex patterns, in particular those occurring in time series, is then carried on in the contribution by Pike, where he presents a rather general formalism. The paper by Guettinger et al. elucidates pattern formation and pattern recognition again from a rather general point of view. We then proceed to physical realizations of abstract models on pattern recognition, where models of spin glasses representing neural nets are an important intermediate step. After the seminal paper of J. Hopfield a number of physicists who are experts in spin glasses and statistical physics have studied such spin glass models in great detail. Recent results are presented by Horner and Kanter in these proceedings. Possible realizations of associative memory designs are discussed by Dreyfus et al. At least at this point the reader of these proceedings may get a feeling which seized all participants of this symposium, namely how fruitful and fascinating an interdisciplinary discussion ranging from basic problems to applications can be.

The next part of these proceedings is devoted to recent studies how to realize microelectronic elements of new types. A device suggested by Mahler et al. bears great promise to act as a microelectronic computer element. The paper by Harrison et al. provides us with an example of the production of chaotic processes in lasers. The instability towards chaos can be considered in two ways : some people consider any instability as a nuisance because it may spoil the proper performance of the laser to be used, for instance, in an all-optical computer. But one may consider laser chaos also from a different point of view, e.g. as a random generator, or we may recall that even random computers have been suggested. The paper by Cerny on statistically coupled processors deals in particular with the question how either ill-defined problems can be tackled in a well-defined manner or how problems can be cast into a variational form which can be solved by coupled processors. It appears to me that some of these approaches can be connected with the idea of a Lagrangian variational principle in the way that the energy function introduced by Cerny is the function to be minimized under the constraints of the given equations. McCauley treats a problem which is of great current interest also, namely to which extent chaotic orbits can be calculated. This is of particular importance because more and more chaotic processes have been discovered and analyzed in natural systems, e.g. fluids, lasers, electronic analog-computers, and it is a fundamental problem in how far such processes can be adequately mimicked on digital computers. The paper by Antoranz et al. deals with an explicit example of chaos, namely intermittency. Finally the contribution by Caglioti et al. makes us acquainted with some typical problems in robotics.

I believe that these proceedings will provide the reader with the various facets which we have to consider when we are dealing with computational processes especially those needed for pattern recognition and associative memory. Clearly, each of these contributions can again give only some cross section of the individual fields. But I think the reader will now be enabled to go on and make his way through the various fields and to synthesize what he has read here and to draw his own picture of the complex world of computation.

The following part of my contribution deals with synergetic computers. To this end I shall first explain some aspects of synergetics we shall need for the construction of our computer model. Eventually I will present an explicit model which will allow us to treat pattern recognition, which is invariant against total intensity, location, orientation, size, and to some extent against deformations of the patterns.

2. A Reminder of Some Basic Problems and Ideas of Synergetics

The interdisciplinary field of synergetics deals with the spontaneous formation of spatial, temporal or functional structures by systems which have many components. A wellknown simple example for the formation of such structures is provided by the convection instability in fluid dynamics (Fig. 2.1). When a fluid is heated from below, beyond a critical temperature difference between the lower and upper surface a specific spatial pattern is formed, namely rolls, in which the fluid moves upwards and downwards, respectively. Thus, even by a homogeneous, i.e. unspecific heating from below the formerly homogeneous or structureless state can acquire a structure or pattern. Another example for the spontaneous formation of patterns is provided by certain chemical reactions. When the reactants are brought together, they may spontaneously form a layered structure, or spirals, or concentric waves which are running outwards (Fig. 2.2).

Let us consider the processes of chemical pattern formation in some more detail (Fig. 2.3). Let us divide the total volume into small volume elements. Then in each of these volume elements chemical processes are going on. As can be shown, such processes can be interpreted as computations, for instance the multiplication of concentrations, the addition of concentrations, etc. Furthermore the individual volume elements are coupled by diffusion, i.e. by the exchange of molecules, or in other words, by the exchange of specific data. So we have here a simple model of a parallel computer. The result of the computation is the evolving macroscopic

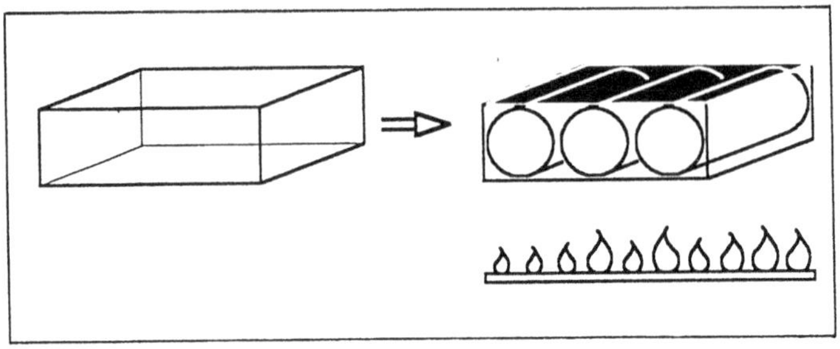

Fig. 2.1

The convection instability

When a fluid is heated from below, it may spontaneously form a roll pattern (r.h.s.).

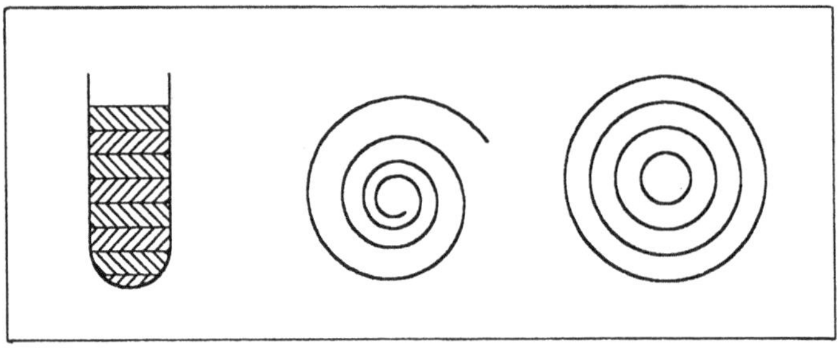

Fig. 2.2

Various patterns formed by chemical reactions, in particular by the Belousov-Zhabotinski reaction. From left to right: stripe patterns, spiral patterns, outgoing circular waves

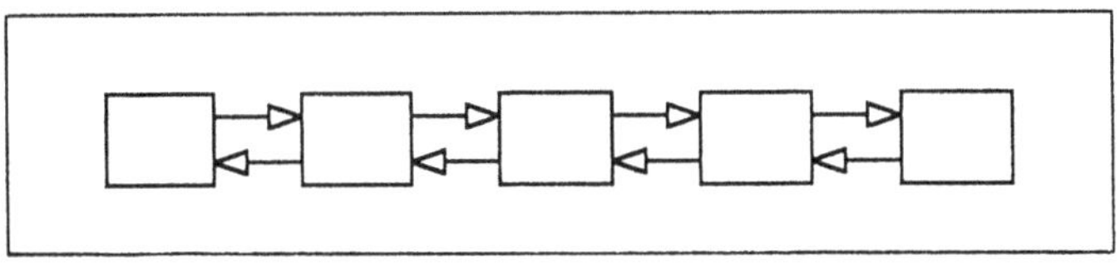

Fig. 2.3

Interpretation of chemical processes as parallel computer

pattern, or in other words, the macroscopic state of the system. This state depends on the control parameters, on the boundary conditions, and on initial conditions. Similarly the pattern formation in fluids may be considered as a result of macroscopic parallel processing.

We now briefly remind the reader of some basic steps which are done to derive such patterns mathematically. What we are interested in are qualitative changes on macroscopic scales, or in other words, the formation of new kinds of structures or patterns. We describe the total system at the microscopic or mesoscopic level by a state vector $\underset{\sim}{q}$

$$\underset{\sim}{q} = \begin{pmatrix} q_1 \\ \vdots \\ q_N \end{pmatrix}, \qquad \underset{\sim}{q} = \underset{\sim}{q}\,(\underset{\sim}{x}, t) \tag{2.1}$$

where t is time and $\underset{\sim}{x}$ denotes the spatial coordinate.
It is assumed that this state vector obeys a set of evolution equations

$$\dot{\underset{\sim}{q}} = \underset{\sim}{N}\,(\underset{\sim}{q}, \alpha) + \underset{\sim}{F} \tag{2.2}$$

where $\underset{\sim}{N}$ is a nonlinear function which may contain spatial derivatives and α is one or a set of control parameters. $\underset{\sim}{F}$ represents fluctuating forces which in general are assumed to be δ-correlated in time and Gaussian distributed. When the control parameter α is changed beyond a critical value, a pattern can become unstable and is replaced by a new one. To check this change, a linear stability analysis is to be performed which, as solution, yields collective modes which can be written as

$$\underset{\sim}{w}(\underset{\sim}{x}, t) = \exp(\lambda t)\, \underset{\sim}{v}(\underset{\sim}{x}). \qquad (2.3)$$

The modes, $\underset{\sim}{v}$, where Re $\lambda \geq 0$ are denoted as unstable modes, $\underset{\sim}{v}_u$, the modes with Re $\lambda > 0$ as the stable modes, $\underset{\sim}{v}_s$. The solution $\underset{\sim}{q}$ of (2.2) is then written as

$$\underset{\sim}{q}(\underset{\sim}{x}, t) = \underset{\sim}{q}_o + \sum_u \xi_u(t)\, \underset{\sim}{v}_u(\underset{\sim}{x}) + \sum_s \xi_s(t)\, \underset{\sim}{v}(\underset{\sim}{x}) \qquad (2.4)$$

where $\underset{\sim}{q}_o$ is, loosely speaking, the solution of (2.2) before the instability happened. The amplitudes $\xi_u(t)$ are called <u>order parameters</u>, whereas $\xi_s(t)$ are the slaved mode amplitudes. According to the slaving principle, the slaved mode amplitudes, ξ_s, are uniquely determined by the order parameters and the fluctuations so that the dynamics is determined by few variables, the order parameters ξ_u. Thus at a transition point an <u>information compression</u> takes place. The evolving order parameters determine specific spatial, spatio-temporal, or functional patterns. The behavior of the system can be discussed either in the space of the state vector $\underset{\sim}{q}$ or in the order parameter space. For the sake of visualization let us consider the order parameter space, which is usually low dimensional. Then the behavior of systems can be classified at least to some degree. The most simple state is that of a fixed point, i.e. a time-independent state to which all neighboring trajectories are attracted (Fig. 2.4). Such a case is in particular realized when the temporal change of $\underset{\sim}{q}$ is determined by the gradient of a potential function V

$$\dot{\underset{\sim}{q}} = -\,\text{grad}_{\underset{\sim}{q}}\, V(\underset{\sim}{q}). \qquad (2.5)$$

V looks like a landscape as depicted in Fig. 2.5. (2.5) is the equation of a particle which moves in an overdamped fashion, i.e. in a heavily damped fashion in a potential V. Quite clearly the attractors are in each case the bottoms of the valleys and the basin of attraction starts at the top of each hill in between which then leads the sliding particle to a specific minimum. When we include fluctuating forces $\underset{\sim}{F}$, the particle is not only subject to the force caused by the potential V, but also to the random pushes exerted on it by the fluctuating forces $\underset{\sim}{F}$

$$\dot{\underset{\sim}{q}} = -\,\text{grad}_{\underset{\sim}{q}}\, V + \underset{\sim}{F}(t). \qquad (2.6)$$

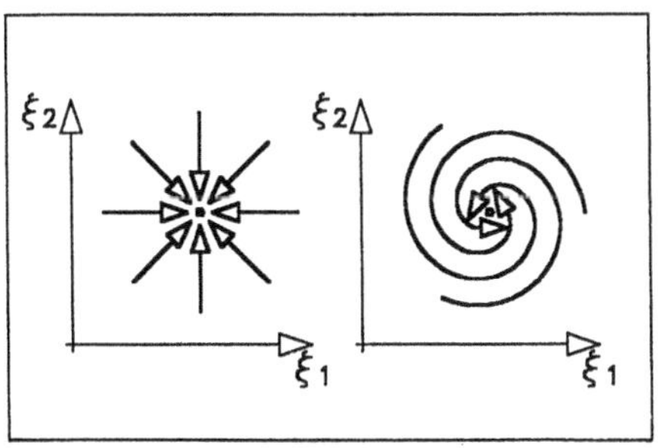

<u>Fig. 2.4</u> Simple example of attractors: fixed points l.h.s. node, r.h.s. focus

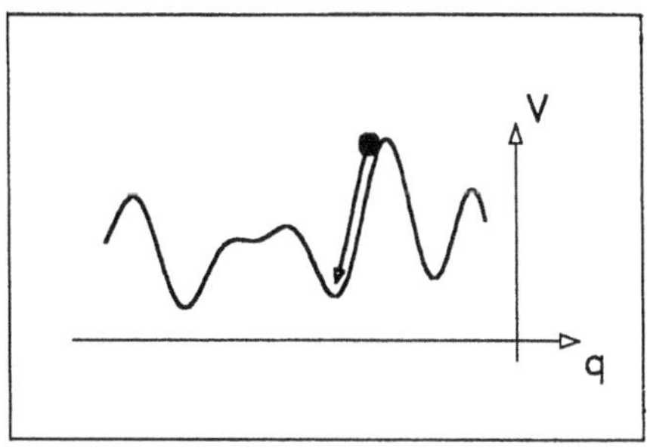

<u>Fig. 2.5</u> Visualization of the overdamped motion of a particle in a potential V

By means of the Fokker-Planck equation one can easily determine the distribution function of the state vector (or of the order parameters), f, according to

$$f\,(\underline{q}) = N \exp(-\,2\,V(\underline{q})/Q) \tag{2.7}$$

where Q represents the strength of the fluctuating forces, which in our case is assumed to be the same for each component of $\underline{F}$. In equilibrium thermodynamics Q is proportional to the temperature. Clearly when we increase Q or the temperature, the particle may jump over hills and therefore can eventually acquire the lowest "energy" state. For more complicated nonlinear functions in equation (2.2), different kinds of behavior can occur. Three typical are : 1) The motion on a limit cycle in $\underline{q}$-space, or in other words, an oscillatory motion, if $\underline{q}$ is considered as a function of time. 2) Other kinds of attractors are provided by socalled tori, where Q is a quasi-periodic function. 3) More recently chaotic attractors have come to the focus of the attention of mathematicians and physicists where the trajectories $\underline{q}$ (t) are time-dependent and stay in a closed region in $\underline{q}$-space, and where originally nearby trajectories separate exponentially. Trajectories outside those chaotic attractors are attracted into these regions.

Synergetics mainly deals with transitions between these different kinds of attractors caused by a change of control parameters. In general when one attractor becomes unstable, not only a single new attractor may occur but a set of them. So the system has a choice in verifying one out of several possibilities with respect to its evolving patterns. It is this ambiguity which we shall employ below to establish a strong analogy between pattern formation as treated here and pattern recognition by synergetic computers.

In conclusion of this section we may state that in a nut shell we presented some of the concepts used in synergetics. We now wish to address ourselves to computational processes, in particular those which serve pattern recognition or associative memory.

3. Associative Memory and Pattern Recognition

Since these words are not always used with the same meaning, we shall dwell a little bit on their proper definition.

a) Associative Memory
Let us consider a pattern or a set of data. Then we understand by associative memory a device or process with the following property: When the pattern or the data are presented only incompletely to the system, it may restore the whole pattern or set of data.

b) Pattern Recognition
When a complete or incomplete pattern or a set of data is presented to a system, it may attach different labels to different patterns, e.g. it may attach names to different faces. There is also the feature of categorisation where patterns offered to a system are put into the same class or different classes, for instance we may have classes such as horses, dogs, cats, etc. The first task of pattern recognition can be reduced to that of associative memory. So when a pattern v is to be transformed into labels f, we form a new data set by (v, f). Then we offer the system v, which, because of being an associative memory restores (v, f). Then eventually we need only to read off f. In this way we have achieved a transformation from v to f.

Let us now discuss some ways how these tasks may be fulfilled. There is by now quite a literature on networks (Fig. 3.1) and, more recently on spin glasses (Fig. 3.2) (cf. also the contributions to this volume), where the Hopfield model has become the prominent one. When thinking of brains we know that their neurons form a network via connections, namely axons and dendrites. To model the behavior of such a neural network, one may attribute only two states to each neuron,

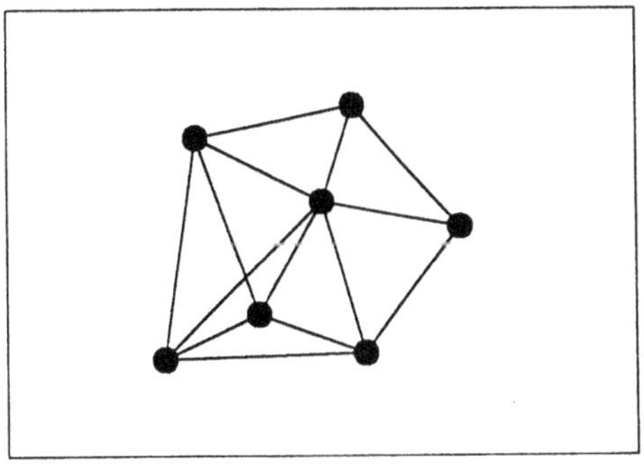

Fig. 3.1 Example of a network

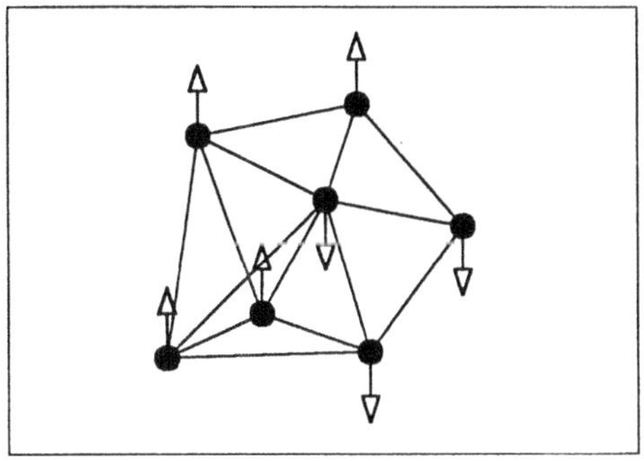

Fig. 3.2 Example of spin glass

according to the McCulloch-Pitts model. With respect to the role of the connections, the concept of Hebb's synapse is still playing an important role. Loosely speaking, according to this idea synapses between neurons are enhanced by their use. In this way learning is achieved by a network. When we combine these two concepts on neurons and their connections we are led in a rather natural way to the spin glass model of a neural network. As we all know, a spin can be visualized as an elementary magnet which can point in two directions only, for instance up or down. Thus a spin can occupy only two states where the state up may be identified with a neuron being active, the state down with a neuron being at rest. The connections can be represented by interaction energies J_{ij} between the spins i, j. In the Hopfield model therefore the total interaction energy is written as

$$E = -\frac{1}{2}\sum_{i,j} J_{i,j}\, s_i\, s_j \,. \tag{3.1}$$

It is now assumed that the system can learn patterns offered to it by means of specific choices of J_{ij}. We shall come back to this question below. In theory one can easily fix the J_{ij} but in practice it is still a problem how to implement learning into a physical device. Because the system has a potential energy, we may assume that it occupies one of its low-lying attractor states as presented by Fig. 2.5. Note however that Fig. 2.5 refers to a continuous variable, whereas here we deal with spin variables, which can acquire the values $s_i = \pm 1$ ($\pm$ 1/2) only. The study of the properties of these spin glasses with respect to learning etc. is a fascinating problem to scientists, who have been working on spin glasses in other contexts previously, and we can refer the reader to some of the most recent developments in this field to articles contained in this volume.

4. Synergetic Computers

In this section I wish to show how associative memory and pattern recognition can be achieved by processes which are strongly reminiscent of pattern formation in synergetic systems. This leads us to consider specific classes of synergetic computers. "Synergetic" will mean as usual that the result of a computation is achieved by the joint, i.e. in particular parallel action of its individual parts. But below we shall become more specific and explicit. While I have pointed to analogies between pattern recognition and pattern formation on previous occasions and some of the underlying processes have been utilized by Shimizu to devise his synergetic (or holonic) computer, my more recent work in this field was triggered by work by my coworker W. Banzhaf. He considered more or less the Hopfield model but now requiring that

$$V = -\,\underset{\sim}{q}\, J\, \underset{\sim}{q}$$

becomes a minimum where $\underset{\sim}{q}$ is normalized. He considered a vector out of a space in which the individual components q_i can aquire only integer numbers in between -5 and +5. His numerical computations then showed that in this way he could restore patterns, i.e. he realized an associative memory.

My own approach consists of several steps which I am going to explain now. In order to visualize my procedure, let us consider a one-dimensional pattern, for instance different intensities in grey along a line. Let us divide this line into intervals of equal length. Then a pattern will be denoted by the vector $\underset{\sim}{u}$ (Fig. 4.1):

$$\underset{\sim}{u} = \begin{pmatrix} u_1 \\ u_N \end{pmatrix} . \tag{4.1}$$

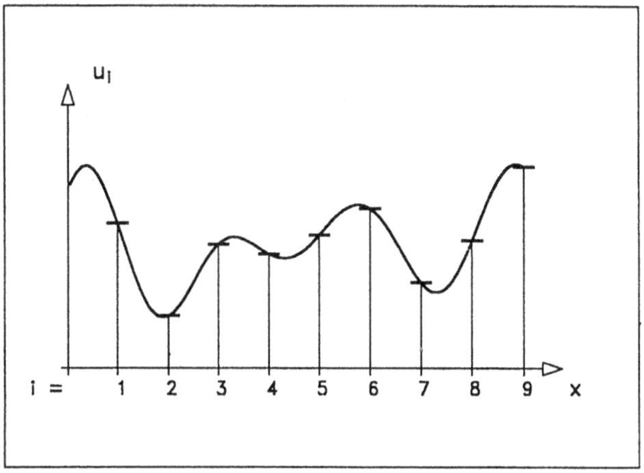

Fig. 4.1

Decomposition of one-dimensional pattern into discrete values

In the following we shall replace each of such a prototype pattern by

$$\underset{\sim}{v}^{(1)} = N\,(\underset{\sim}{u} - \bar{\underset{\sim}{u}}) \tag{4.2}$$

where the mean value $\bar{\underset{\sim}{u}}$ is defined by

$$\bar{\underset{\sim}{u}} = \underset{\sim}{1} \cdot \sum_j u_j \ . \tag{4.3}$$

The factor N serves for normalization so that

$$\underset{\sim}{v}^{(1)+} \underset{\sim}{v}^{(1)} = 1 \ . \tag{4.4}$$

$\underset{\sim}{v}^{(1)+}$ is the adjoint vector to $\underset{\sim}{v}^{(1)}$. We then define the learning matrix J as usual by

$$2\,J = -\lambda\,\underset{\sim}{v}^{(1)} \bullet \underset{\sim}{v}^{(1)+} \tag{4.5}$$

if only a single prototype pattern is present. Note the sequence of the two vectors $\underset{\sim}{v}^{(1)}$ and $\underset{\sim}{v}^{(1)+}$ which means that we have to take first a scalar product between $\underset{\sim}{v}^{(1)+}$ and a vector to be multiplied from the right-hand side and then we have to multiply this scalar by $\tilde{v}^{(1)}$. In particular we may form a potential energy function V by means of

$$V = \underset{\sim}{q}^{+} J \underset{\sim}{q} . \tag{4.6}$$

We might now try to restore a pattern in the following way: We start from some incorrect or incomplete vector $\underset{\sim}{q}$ and then we try to minimize V by proper choices of $\underset{\sim}{q}$. Such a procedure is followed up for instance in socalled evolution strategies. Among those is the one which can be derived from our considerations of section 3, namely we interpret V as the potential energy and we assume that a particle moves in an overdamped fashion within such a potential according to

$$\dot{\underset{\sim}{q}} = - \operatorname{grad}_{\underset{\sim}{q}} V = 2 J \underset{\sim}{q} . \tag{4.7}$$

In the next step we shall assume that different prototype patterns $\tilde{u}^{(k)}$, k=1,.., M, are learned. Then in a first step we form (4.2,3)

$$\tilde{v}^{(k)} = N_k (\tilde{u}^{(k)} - \bar{\tilde{u}}^{(k)}) \tag{4.8}$$

where the normalization factors N_k are chosen such that

$$\tilde{v}^{(k)+} \tilde{v}^{(k')} = \delta_{kk'} \tag{4.9}$$

holds. In addition it was assumed that the prototype patterns V are orthogonal to each other. Then as usual the learning matrix is defined by

$$2 J = \sum_{k=1}^{M} \lambda_k \tilde{v}^{(k)} \bullet \tilde{v}^{(k)+} . \tag{4.10}$$

Let us now study the meaning of (4.7). To this end we insert the hypothesis

$$\underset{\sim}{q} = e^{\lambda t} \tilde{w} \tag{4.11}$$

into (4.7), which yields

$$\lambda \tilde{w} = \sum_{k=1}^{M} \lambda_k \tilde{v}^{(k)} \bullet \tilde{v}^{(k)+} \tilde{w} . \tag{4.12}$$

Because of the relation (4.9) it is easy to convince oneself that the eigen-vectors $\tilde{w}$ of the equation (4.12) are given by

$$\tilde{w}^{(k)} = \tilde{v}^{(k)}, \; k = 1, \ldots, M \tag{4.13}$$

with eigenvalues

$$\lambda = \lambda_k > 0 . \tag{4.14}$$

In case M ≤ N the eigenvectors must be supplemented by the further set

$$\tilde{w}^{(k')} \perp \tilde{v}^{(k)} \qquad k<k' \leq N \tag{4.15}$$

with eigenvalues

$$\lambda = \lambda_k = 0 . \tag{4.16}$$

Thus the general solution of (4.7) may be written as

$$\underset{\sim}{q} = \sum_{k=1}^{M} c_k \exp(\lambda_k t)\, \underset{\sim}{v}^{(k)} + \sum_{k'=M+1}^{N} c_{k'}\, \underset{\sim}{v}^{(k')} \qquad (4.17)$$

where the coefficients c_k are given by

$$c_k = \underset{\sim}{w}^{(k)+}\, \underset{\sim}{q}(0) \qquad (4.18)$$

where $\underset{\sim}{q}(0)$ is the initial value of $\underset{\sim}{q}$ which need not coincide with one of the prototype patterns. Rather it is the task of the procedure to identify $\underset{\sim}{q}$ with one of the prototype patterns, i.e. either to correct the errors in $\underset{\sim}{q}$ or to restore it. In case that

$$\lambda_k = \lambda \qquad (4.19)$$

holds, we readily find

$$\underset{\sim}{q}^{(t)} \underset{t\to\infty}{\to} e^{\lambda t} \sum_{k=1}^{M} c_k\, \underset{\sim}{v}^{(k)} + \text{const.} \qquad (4.20)$$

The advantages and drawbacks of (4.20) can best be illustrated by two typical examples. Let us consider a two-dimensional space in which there is only one prototype pattern $\underset{\sim}{v}^{(1)}$, and the second vector is, so to speak, a pattern not offered to the system (compare Fig. 4.2). Then in the course of time $\underset{\sim}{q}$ will increase in a direction parallel to $\underset{\sim}{v}_1$, i.e. $\underset{\sim}{q}$ will become more and more parallel to $\underset{\sim}{v}_1$.

Thus when we keep $\underset{\sim}{q}(t)$ normalized, $\underset{\sim}{q}$ becomes more and more identical with the original prototype pattern. So our approach seems to be quite satisfactory. But now consider the case of two prototype patterns (compare Fig. 4.3). According to equation (4.7) with (4.10) we have now two unstable directions. Therefore, $|\underset{\sim}{q}|$ will increase, having all the time the same angle with respect to the $\underset{\sim}{v}^{(1)}$ axis, i.e. no distinction between $\underset{\sim}{v}^{(1)}$ and $\underset{\sim}{v}^{(2)}$ is achieved. In order to achieve such a distinction it suggests itself to introduce a potential function V_1 which pulls $\underset{\sim}{q}$ closer to $\underset{\sim}{v}^{(1)}$ provided it lays within an angle $\phi < 45^0$ or to $\underset{\sim}{v}^{(2)}$ provided this angle is in between 45^0 and 90^0. Thus we require a potential function V_1 as depicted in

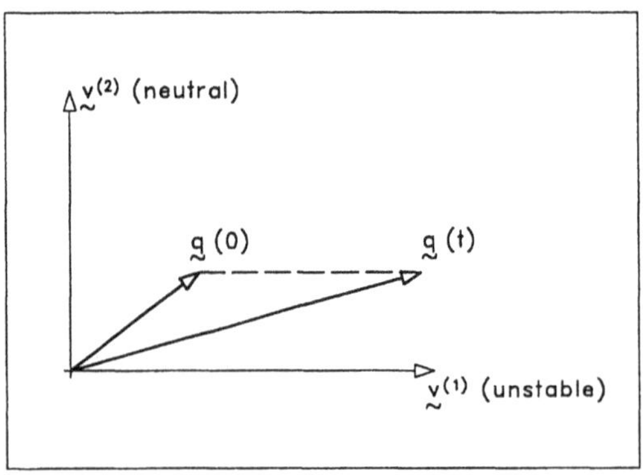

Fig. 4.2 See text

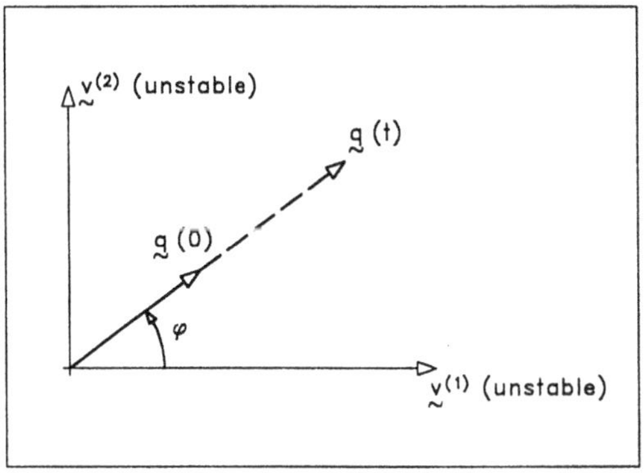

Fig. 4.3. See text

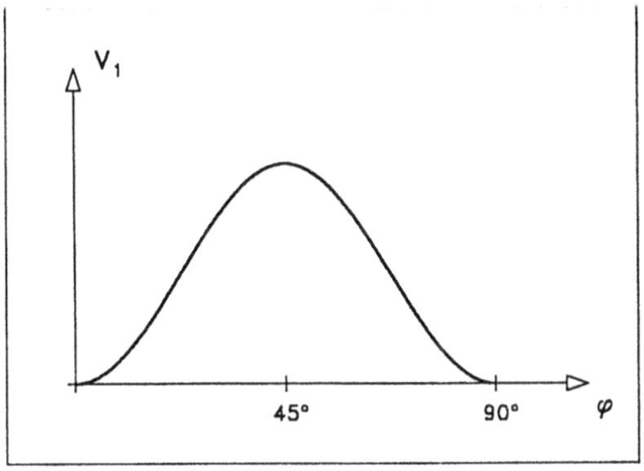

Fig. 4.4

The potential required to pull the offered state vector into the desired direction

Fig. 4.4. Such a potential function can be easily constructed by making the following hypothesis

$$V_1 = C_1\, (\underset{\sim}{v}^{(1)+}\underset{\sim}{q})^2\, (\underset{\sim}{v}^{(2)+}\underset{\sim}{q})^2 \,. \qquad (4.21)$$

A simple calculation shows that (4.21) can be expressed as

$$C_1\, q^4\, \frac{1}{2} \sin^2 (2\phi) \,. \qquad (4.22)$$

The equation of motion is then given by

$$\dot{\underset{\sim}{q}} = 2\, J\, \underset{\sim}{q} - \mathrm{grad}_{\underset{\sim}{q}}\, V_1 \,. \qquad (4.23)$$

One may readily see that the absolute value of $\underset{\sim}{q}$ will increase for ever even in the model (4.23). But because we wish that the system is pulled into an attractor we introduce a saturation in the form of the potential

$$V_2\, (\underset{\sim}{q}) = C_2\, |\underset{\sim}{q}|^4 \,. \qquad (4.24)$$

We are now in a position to formulate our model in its present form. We have the equation

$$\dot{\underset{\sim}{q}} = -\, \mathrm{grad}_{\underset{\sim}{q}}\, V \qquad (4.25)$$

where the potential V is given by

$$V = V_o + V_1 + V_2 . \tag{4.26}$$

Here, in the general multicomponent case the potentials are given by

$$V_o = - \underset{\sim}{q}^+ \sum_{k=1}^{M} \lambda_k \, \underset{\sim}{v}^{(k)} \bullet \underset{\sim}{v}^{(k)+} \, \underset{\sim}{q} \tag{4.27}$$

where

$$\lambda_k > 0 \tag{4.27a}$$

$$V_1 = C_1 \sum_{k \neq k'}^{M} (\underset{\sim}{v}^{(k)+} \underset{\sim}{q})^2 \, (\underset{\sim}{v}^{(k')+} \underset{\sim}{q})^2 \tag{4.28}$$

and (4.24). (4.28) can easily be generalized by introducing instead of the constant C_1 the constants $C_{1,k,k'}$. One readily convinces oneself that V_1 has its minimum, namely = 0, if

$$\underset{\sim}{q} = v^{(k_o)} \tag{4.29}$$

holds, because then all other products $\underset{\sim}{q}^+ \bullet \underset{\sim}{v}^{(k)}$, $k \neq k_o$, vanish. This lets us expect that our procedure is capable of bringing $\underset{\sim}{q}$ into any one of the prototype patterns provided $\underset{\sim}{q}$ lies initially close enough to the corresponding axis. The equation (4.25) reads explicitly

$$\dot{\underset{\sim}{q}} = 2 \, J \, \underset{\sim}{q} - 2C_1 \sum_{\underset{\sim}{k} \neq \underset{\sim}{k}'}^{M} (\underset{\sim}{v}^{(k)+} \underset{\sim}{q})^2 \, (\underset{\sim}{v}^{(k')+} \underset{\sim}{q}) \, \underset{\sim}{v}^{(k')}$$

$$- 4 \, C_2 \, \underset{\sim}{q}^2 \, \underset{\sim}{q} \, . \tag{4.30}$$

In the following we shall assume that the pattern $\underset{\sim}{q}$, which is offered to the system, is preprocessed so that its average value is 0 (cf. (4.2))In general the number M of prototype patterns will be smaller than the dimension N of the space. In such a case <u>any</u> pattern presented to the system as an initial state $\underset{\sim}{q}(0)$ will be "pulled" into the space of prototype patterns, provided $\underset{\sim}{q}(0)$ has some component in that space. In order to avoid that $\underset{\sim}{q}$'s with small components are identified as some pattern k, we must require that $\underset{\sim}{q}(0)$ lies sufficiently close to the M-space. To this end we define

$$\underset{\sim}{q}_{||} = \sum_{k}^{M} \underset{\sim}{v}^{(k)} \, \underset{\sim}{v}^{(k)+} \, \underset{\sim}{q} \, . \tag{4.31}$$

This allows us to form

$$\underset{\sim}{q}_{||}^2 = \sum_{k}^{M} d_k^2 \tag{4.32}$$

where

$$d_k = \underset{\sim}{v}^{(k)+} \underset{\sim}{q} \tag{4.33}$$

We now admit only those $\underset{\sim}{q}$ for which

$$\underset{\sim}{q}_{||}^2 > a \, \underset{\sim}{q}^2 . \tag{4.34}$$

A typical choice of a is

$$a = \frac{1}{2} . \tag{4.35}$$

If

$$\underset{\sim}{q}_{||}^2 = \underset{\sim}{q} \qquad (4.36)$$

then $\underset{\sim}{q}(0)$ lies in the M-space from the very beginning.

We now mention some of the properties of this model, which can easily be checked. The following statements can be made. The extrema of V, in particular the minima, can be determined explicitly. The deepest minimum is realized when only one d_k is unequal to 0 and all others vanish. The depth of V can be determined. If the dimension of the space is N, N patterns can be stored and identified. When we introduce the new variables (4.33) the equations (4.30) acquire the form

$$\dot{d}_k = \lambda\, d_k - 4C_1 \sum_{k' \neq k} d_k d_{k'}^2 - 4C_2 \sum_{k'} d_{k'}^2 d_k, \quad k=1,\ldots,M \qquad (4.37)$$

Quite similar equations occur in fluid dynamics and in specific types of multimode lasers. Let us discuss the relationship with fluid dynamics more closely. To this end we replace the indices of the vector components ,j, by a continuous variable

$$j \to \underset{\sim}{x} = (x,y)\,. \qquad (4.38)$$

We further identify $\underset{\sim}{v}$ with a plane wave

$$\underset{\sim}{v}^{(k)} \to \exp(i\underset{\sim}{k}\underset{\sim}{x}), \quad \underset{\sim}{q} \to q(\underset{\sim}{x})\,. \qquad (4.39)$$

The sum in $\underset{\sim}{v}^{(k)+}\underset{\sim}{q} = \sum_j v_j^{(k)+} q_j$

then becomes an integral

$$\underset{\sim}{v}^{(k)+}\underset{\sim}{q} \Rightarrow \int e^{-i\underset{\sim}{k}\underset{\sim}{x}}\, q(\underset{\sim}{x})\, d^2x = d_k\,. \qquad (4.40)$$

For the absolute value of $\underset{\sim}{k}$ fixed, (4.39) represents a set of role patterns in fluids with different orientations. The equations (4.37) are identical with those of the formation of role patterns in fluid dynamics within a specific model. We know from there that only one specific set of parallel roles can exist and that this set is determined by initial conditions or initial fluctuations. But such initial fluctuations mean that originally a fuzzy pattern was offered to the liquid, which then recognizes one of its inherent patterns (4.34) by means of choosing a specific state $d_k \neq 0$, all others = 0. Note that the prototype patterns $\underset{\sim}{v}^{(k)}$ form the basis of the order parameters of the system. Quite clearly here we have found a physical system, which by means of its synergism can reproduce or restore a pattern and thus acts as an associative memory or as pattern recognizer. Things will become still more attractive if such systems can not only be realized by fluids, but also by lasers, so that these considerations may be of relevance to all-optical computers.

In conclusion of this section we may state that our present model has some strikingly simple features. We may immediately determine the total number of patterns which can be stored, we may immediately overlook the basins of attraction, we may calculate the depths of the steady-state potentials and we may calculate relaxation times close to attracting states. In the spirit of section 2 we may add fluctuating forces to facilitate the jumping of the system of a metastable state into more stable states. One can then easily calculate the distribution function in the form (2.7). When we consider λ as a parameter which may be changed from negative to positive values, we observe all the features of a nonequilibrium phase transition, discussed in extenso in the literature on synergetics.

To enhance the speed of computations, (4.37) can be transformed into an equation for

$$n_k = d_k^2 \tag{4.41}$$

namely

$$2\dot{n}_k = \lambda n_k - 4C_1 \sum_{k' \neq k} n_{k'} n_k - 4C_2 \sum_{k'} n_{k'} n_k \,. \tag{4.42}$$

Our model allows for a number of generalizations and we wish to discuss a few of them in the next paragraph.

5. Some Generalizations of the Simple Model of Section 4

First of all we may assume that the values λ become different, λ_k. Then one can easily show that the deepest minimum is the one with the biggest λ_k, provided the ratio of $C_1 : C_2$ is big enough. If we interpret λ_k as an attention parameter, then that pattern is identified to which most attention is being paid. On the other hand, the "attention-parameter" λ_k may decrease due to neural saturation effects. To model this we may introduce λ_k as a variable which in turn depends on the "occupation number" d_k^2. Thus after $\underset{\sim}{q}$ has reached a specific attractor d_k, the corresponding λ_k may be lowered (even to negative values) so that this (former) attractor is destabilized. Thus, if the initial pattern $\underset{\sim}{q}(0)$ is offered anew to the system, $\underset{\sim}{q}$ will reach another attractor with $d_{k'}$, the next best. Let us further assume that λ_k recovers after a while. Then, when we offer the system an initial pattern which has practically the same overlap with, say $\underset{\sim}{v}^{(k)}$ and $\underset{\sim}{v}^{(k')}$, the following will happen. First the system will go to the $\underset{\sim}{v}^{(k)}$ - axis ("d_k"), then to the $\underset{\sim}{v}^{(k')}$ - axis ("$d_{k'}$"),but because $d_{k'}$ saturates, it will return to d_k and so on. In other words, the system will oscillate between the states d_k and $d_{k'}$. Such oscillatory behavior is indeed observed when looking at ambiguous patterns, e.g. the famous vase-face transition.

In the present context the biological meaning of this effect becomes clear: It allows the system to perceive <u>all</u> possibilities inherent in the pattern offered to it with respect to the set of stored prototype patterns.

An important generalization being also of great practical interest is that the prototype patterns are no more orthogonal to each other. Consider a set of prototype patterns $\tilde{\underset{\sim}{v}}^{(k)}$ being linearly independent. Then we may always find a nonsingular matrix, Ω or actually a whole set of such matrices so that

$$\Omega^{-1} \tilde{\underset{\sim}{v}}^{(k)} = \underset{\sim}{v}^{(k)} \tag{5.1}$$

is a set of orthogonal vectors. Let us assume that we have transformed $\tilde{\underset{\sim}{q}}$ and $\tilde{\underset{\sim}{v}}$ to this orthogonal set. Then the equations are given as before by

$$\begin{aligned} \dot{\underset{\sim}{q}} = {} & \sum \lambda_k \underset{\sim}{v}^{(k)} \bullet \underset{\sim}{v}^{(k)+} \underset{\sim}{q} \\ & - 2C_1 \sum_{k \neq k'}^{M} (\underset{\sim}{v}^{(k)+} \underset{\sim}{q})^2 (\underset{\sim}{v}^{(k')+} \underset{\sim}{q}) \, \underset{\sim}{v}^{(k')} \\ & - 4C_2 \, \underset{\sim}{q}^2 \, \underset{\sim}{q} \,. \end{aligned} \tag{5.2}$$

Making use of the transformations

$$\underset{\sim}{v}^{(k)} = \Omega \, \tilde{\underset{\sim}{v}}^{(k)} \tag{5.3}$$

and

$$\underset{\sim}{q}^{(k)} = \Omega \, \underset{\sim}{\tilde{q}} \tag{5.4}$$

we may readily cast (5.2) into the form

$$\begin{aligned}\underset{\sim}{\tilde{q}} = & \sum_k^M \lambda_k \, \underset{\sim}{\tilde{v}}^{(k)} \bullet \underset{\sim}{\tilde{v}}^{(k)+} \Lambda \, \underset{\sim}{\tilde{q}} \\ & - 2C_1 \sum_{k \neq k'}^M \left(\underset{\sim}{\tilde{v}}^{(k)+} \Lambda \, \underset{\sim}{\tilde{q}}\right)^2 \left(\underset{\sim}{v}^{(k')+} \Lambda \, \underset{\sim}{\tilde{q}}\right) \underset{\sim}{\tilde{v}}^{(k')} \\ & - 4C_2 \left(\underset{\sim}{\tilde{q}}^{+} \Lambda \, \underset{\sim}{\tilde{q}}\right) \underset{\sim}{\tilde{q}}\end{aligned} \tag{5.5}$$

where we have used the abbreviation

$$\Omega^{+}\Omega = \Lambda \, . \tag{5.6}$$

Thus we can immediately start from nonorthogonal prototype patterns and vectors $\underset{\sim}{\tilde{q}}$. In practical applications the attractor states may be replaced by outlets ("sinks").

For a physicist it might be tempting to consider also quantized versions in which $\underset{\sim}{q}$ or the corresponding field function $q(\underset{\sim}{x})$ is replaced by operators according to Fermi-Dirac, or Bose-Einstein statistics. We shall not dwell on these aspects here, however.

6. Interpretation as Networks - Grand-mother cell revisited

To this end let us consider the simple relation for the components q_j of a vector $\underset{\sim}{q}$,

$$\dot{q}_j = \sum_k J_{jk} \, q_k \, . \tag{6.1}$$

When we identify q_j as the activity of a cell or neuron with index j we may interpret (6.1) in the following manner : the neuron j changes its activity because of the input it gets from all the other neurons with indices k (Fig. 6.1). Similarly the relationship

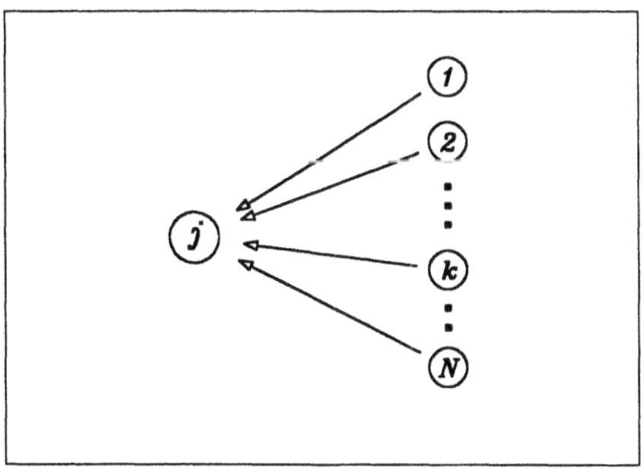

Fig. 6.1 Input of "neurons" 1,2,... N into neuron j

$$\dot{q}_j = \Sigma\, C_{jklm}\, q_k\, q_l\, q_m \qquad (6.2)$$

can be interpreted. But the computation of the inputs has become somewhat more involved. When we assume that the initial values of $\tilde{q}(0)$ are fed into the system from corresponding neurons, we obtain Fig. 6.2 as representation of the network. Evidently there are N connections from the input to the computer neurons, whereas among the latter there are N(N-1) /2 symmetric connections. When we now use the equations (4.37) , we may attribute a network to them, which is depicted in Fig. 6.3.The number of cells can still be reduced when we utilize a model of M < N prototype patterns and the orthogonal space. The properties of orthogonal space are then represented by a neuron with d_0. Here we have N • (M+1) connections from the input to the individual cells, but the cells are connected only by (M+1) • M/2 connections. That means when the number of prototype patterns M is smaller than the number of original elements, we find a considerable reduction of the connections. Finally, we may lump the action of d_k^2 together into their total sum D. This means in other words that we now have introduced a field D and again the number of connections is reduced (compare Fig. 6.4). The total number of connections is now given by N(M+1)+(M+1). The neurons with activities d_j can be considered as grandmother cells, each of which recognizes a specific pattern. How the patterns are identified has become a property of the matrix connecting the initial values $\tilde{q}(0)$ with the d's. In a way we have here found a specific kind of retinotopy if one wishes to interpret the result in this manner. In spite of the fact that the neuron with activities d_k can be considered as a grandmother cell, it must clearly be stated that the

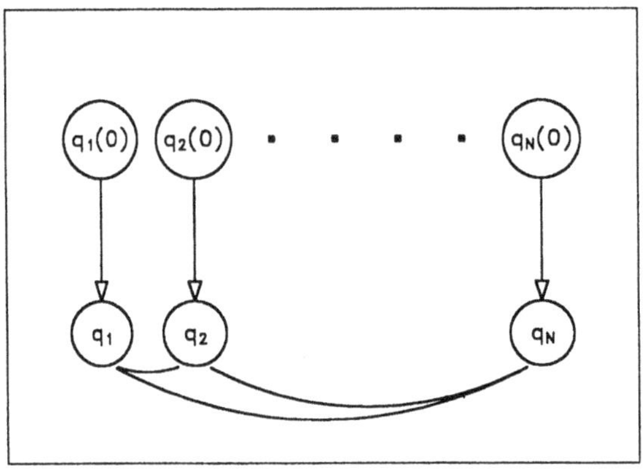

Fig. 6.2

Visualization of the inputs and the neurons with activities $q_1, q_2 \ldots$

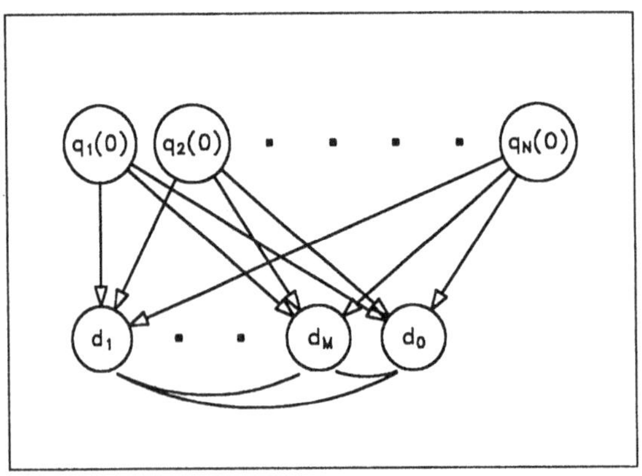

Fig. 6.3

Visualisatuon of grandmother cells with activities $d_1 \ldots d_M$. d_0 takes care of the vector orthogonal to $\tilde{q}_{||}$ (0). Note that the network acts as a whole

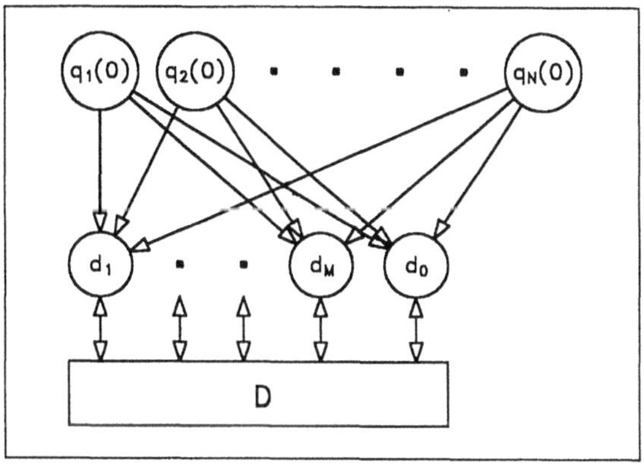

Fig. 6.4

The direct interaction between the cells denoted by $d_1 \ldots d_M$ is here replaced by the interaction with a common field D

identification of a pattern is achieved by the total network, which can be shown by detailed discussion of the equations for d_k. What is new in the present context is in particular the existence of a specific field D. This might be also of some physiological interest. We recognize that for neurons to communicate and to perform computations it is not necessary that they are directly interconnected. Rather they can communicate also by a common field which may be realized for instance by concentrations of specific bio-active molecules. Actually the role of hormones in the body is not so much different, because they also occur in a delocalized fashion. It might be an interesting suggestion, however, to look for similar transmitters and effects in neural networks. In a way the field acts to help to make majority decisions.

7. Recognition invariant against displacements, rotations, scales, deformations and spatial projections

In this section we wish to generalize our above approach so that it can be applied to the problems mentioned in the headings of this section. To this end we introduce the following potential functions. The potential function $\tilde{V}_0$ serves to draw the system into a state with a normalized intensity:

$$\tilde{V}_0 = -\frac{\lambda}{2} \underset{\sim}{q}^2 + C_2 \underset{\sim}{q}^4 \tag{7.1}$$

The next potential function

$$\tilde{V}_1 = f_1 (|\underset{\sim}{q}|) (1 - \sum_k^M d_k^2 / \underset{\sim}{q}^2) \tag{7.2}$$

serves to pull any state vector $\underset{\sim}{q}$ into the space of the prototype patterns. It is assumed that

$$f_1 > 0 \tag{7.3}$$

holds. We have used the definitions

$$d_k = \underset{\sim}{v}^{(k)+} \underset{\sim}{q} \equiv \underset{\sim}{q}^+ \underset{\sim}{v}^{(k)} . \tag{7.4}$$

A convenient choice for f_1 is

$$f_1 = C_1 \underset{\sim}{q}^2 . \tag{7.5}$$

The potential function

$$\tilde{V}_2 = f_2 (|\underset{\sim}{q}|) \sum_{k \neq k'}^{M} d_k^2 d_{k'}^2 / |\underset{\sim}{q}|^4 \tag{7.6}$$

serves for the discrimination of individual modes as was shown above. It will be assumed that

$$f_2 > 0 \tag{7.7}$$

holds. A simple choice is provided by

$$f_2 = C_2 \underset{\sim}{q}^4 . \tag{7.8}$$

The property of (7.1) is obvious from section 4. The minimum of $\tilde{V}_o$ is reached for

$$|\underset{\sim}{q}| = \frac{\lambda}{4C} . \tag{7.9}$$

To discuss the property of (7.2), we note that in general

$$\sum_{k}^{M} d_k^2 = \underset{\sim}{q}_{||}^2 \leq \underset{\sim}{q}^2 . \tag{7.10}$$

As a consequence we obtain

$$\text{Min } \tilde{V}_1 = 0 \text{ for any } \underset{\sim}{q}_{||}^2 = \underset{\sim}{q}^2 . \tag{7.11}$$

The properties of (7.6) were discussed before, we readily find

$$\text{Min } \tilde{V}_2 = 0 \text{ for } d_{k_o} \neq 0, \text{ all others} = 0 \tag{7.12}$$

i.e.

$$\underset{\sim}{q} \; || \; \underset{\sim}{v}^{(k)} \tag{7.13}$$

where the index k denotes a specific prototype pattern. The criterion under which $\underset{\sim}{q}$ will be admitted will be discussed below. To solve the problems indicated in the headings, we shall assume that we are dealing with continuous variables, as indicated above (4.38-40). With respect to any prototype pattern, a pattern offered may be displaced

$$q (\underset{\sim}{x} - \underset{\sim}{\xi}), \tag{7.14}$$

rotated

$$q (R \underset{\sim}{x}), \tag{7.15}$$

having a different size, that is the different scale α

$$q (\alpha \underset{\sim}{x}), \tag{7.16}$$

or quite generally, we may have any coordinate transformation

$$q (S \underset{\sim}{x}) . \tag{7.17}$$

In the following we shall assume that S^{-1} exists. Because q appears always under a scalar product in the potential function, which now has become an integral, we may readily shift the transformation S occurring in q to one occurring in v according to

$$\int v(\underset{\sim}{x})\ q\ (S\underset{\sim}{x})\ d\tau = \int v(S^{-1}\underset{\sim}{x})\ \underset{\sim}{q}\ (\underset{\sim}{x})\ d\tau_s\ . \tag{7.18}$$

The transformation (7.17) is just a special case of

$$w(\underset{\sim}{x}) = \int K(\underset{\sim}{x},\ \underset{\sim}{x}')\ v(\underset{\sim}{x}')\ d\tau\ . \tag{7.19}$$

For instance, when we choose the kernel K in the form

$$K = \delta(x - Sx') \tag{7.20}$$

we readily obtain

$$w(\underset{\sim}{x}) = v(S^{-1}\underset{\sim}{x})\ \rho(S^{-1}\ \underset{\sim}{x})\ . \tag{7.21}$$

In the following we shall assume that K is parametrized by $\underset{\sim}{\eta}$,

$$K(\underset{\sim}{x},\ \underset{\sim}{x}') = K(\underset{\sim}{x},\ \underset{\sim}{x}';\ \underset{\sim}{\eta}) \tag{7.22}$$

where $\underset{\sim}{\eta}$ for instance denotes displacements or rotation, angles, etc. Note that in the transformation (7.19), we may also include projections from a three- to a two-dimensional space. In the following we shall assume that K has the property that

$$\int w_{\underset{\sim}{\eta}}^{(k)}\ (\underset{\sim}{x})\ w_{\underset{\sim}{\eta}}^{(k')}\ (\underset{\sim}{x}) = \delta_{kk'} \tag{7.23}$$

holds. But as we have seen above, our formalism may also be extended to nonorthogonal basis vectors v or w. Because of the parameter $\underset{\sim}{\eta}$, the potential V becomes a function of these variables

$$V = V_o\ (\underset{\sim}{q},\ \underset{\sim}{\eta}) + V_1\ (\underset{\sim}{q},\ \underset{\sim}{\eta}) + V_2\ (\underset{\sim}{q},\ \underset{\sim}{\eta})\ . \tag{7.24}$$

When we now seek for a strategy which minimizes V, we may again resort to equations of the form

$$\dot{\underset{\sim}{q}} = -\ \frac{\partial V}{\partial \underset{\sim}{q}} + F_1(t) \tag{7.25}$$

to which we have now to add equations of the form

$$\dot{\underset{\sim}{\eta}} = -\ \frac{\partial V}{\partial \underset{\sim}{\eta}} + \underset{\sim}{F}_2\ (t) \tag{7.26}$$

however. F_j represents fluctuating forces which will enhance the success of a search strategy to find the minima of V. To enhance the search strategy further, it is helpful to think of the meaning of the variables q and $\underset{\sim}{\eta}$. q represents the pattern offered, whereas $\underset{\sim}{\eta}$ is a variable which serves to match the offered pattern by means of spatial shifts, rotations, etc. Therefore, in order to find a rough agreement, it will be important to match the prototype pattern to the presented pattern quickly, i.e. we shall generalize (7.26) to

$$\dot{\underset{\sim}{\eta}} = -\ \gamma_2\ \frac{\partial V}{\partial \underset{\sim}{\eta}} + \underset{\sim}{F}_2 \tag{7.27}$$

where γ_2 is a time-constant which secures a fast relaxation. On the other hand we shall generalize (7.25) to

$$\dot{\underset{\sim}{q}} = - \gamma_1 \frac{\partial V}{\partial \underset{\sim}{q}} + \underset{\sim}{F}_1 \tag{7.28}$$

so that the pattern q can eventually adjust to the prototype pattern offered. Clearly, when q is a precise pattern, we may ignore (7.25) or (7.28) altogether. Otherwise they will help us to find a perfect match. Because of the construction of our model, any presented pattern, which has an even small component in the prototype space, will be eventually pulled into that space. Because of the presence of noise, any pattern offered would thus be pulled into the prototype space. In order to avoid this difficulty, we must impose a condition for the admission of a vector q, which in generalization of our former condition reads

$$\underset{\underset{\sim}{\eta}}{\mathrm{Max}} \; (\underset{\sim}{q}_{||} / \underset{\sim}{q}^2) > a \tag{7.29}$$

where

$$\underset{\sim}{q}_{||}^2 = \sum_k^M d_k^2 (\underset{\sim}{\eta}) \tag{7.30}$$

is now a function of the parameters $\underset{\sim}{\eta}$. Because q^2 does not depend on $\underset{\sim}{\eta}$, we may replace (7.29) by

$$\underset{\underset{\sim}{\eta}}{\mathrm{Max}} \underset{\sim}{q}_{||}^2 > a \, \underset{\sim}{q}^2 . \tag{7.31}$$

Our above search program for the minima of V provides us, at the same time, with the maximum of left-hand-side of (7.29), but unfortunately only after the whole calculation was done, because (7.29) or (7.31) are not requirements a priori but rather self-consistency requirements, which can be imposed on the procedure only after the solution was found. Our procedure allows for a number of generalizations. For instance we may include colour codings by making the replacements

$$v(\underset{\sim}{x}) \to \underset{\sim}{v}(\underset{\sim}{x}), \qquad v_i(\underset{\sim}{x}) \quad i \to \text{colour}, \tag{7.32}$$

or when colour is treated as a continuous variable by the corresponding generalization

$$v(\underset{\sim}{x}) \to v(\underset{\sim}{x}, \omega) \tag{7.33}$$

where ω is the frequency.

In conclusion we mention that prototype patterns and the pattern to be recognized may be preprocessed, e.g. by making the replacements

$$v(\underset{\sim}{x}) \to \mathrm{grad}\; v(\underset{\sim}{x}) \tag{7.34}$$

$$q(\underset{\sim}{x}) \to \mathrm{grad}\; q(\underset{\sim}{x}) . \tag{7.35}$$

8. Pattern Recognition in Composite Scenes

Here I wish to show how our general algorithm can be applied to more complicated tasks, such as to identify partially hidden patterns.
Consider Fig. 8.0. The solution depends on the task.

Fig. 8.0 Composite scene

Fig. 8.1 Face 1

Fig. 8.2 Face 2

1) We have stored Fig. 8.1 as prototype pattern and wish to recognize 8.1.
2) We have stored Fig. 8.2 as prototype pattern and wish to recognize 8.2.
3) We have stored Figs. 8.1 and 8.2 as prototype patterns and wish to recognize both 8.1 and 8.2.

Solution:

1) We choose $\lambda_1>0$ large, all others <0, and treat 8.0 as the blurred pattern of 8.1.
2) We choose $\lambda_2>0$ large, all others <0, and treat 8.0 as the blurred pattern of 8.2. If 8.1 and 8.2 are stored, also the following solution is possible.
3) First step: We choose $\lambda_1 = \lambda_2>0$, all others <0.
   8.1 will be identified because of majority decision (more common features).
   Second step: We choose $\lambda_1<0$, $\lambda_2>0$, all others <0
   and replace $\underset{\sim}{q}(0)$ by

$$\underset{\sim}{q}(0) = \underset{\sim}{q}_{\xi_0}(0) - \alpha \underset{\sim}{v}^{(1)}$$

so that $\underset{\sim}{v}^{(1)+}\tilde{\underset{\sim}{q}} = 0$.

Then search again (fix new ξ).
Now 8.2 will be identified.

Of course, if 8.0 contains faces different from 8.1, 8.2, no such faces will be identified.

9. Concluding Remarks

In the main part of my above paper I have tried to present some simple models for the realization of synergetic computers. I believe that this kind of computer shows some promise, in particular with respect to all-optical computers, because here we need not consider the digitized problem; rather, for instance, field intensities may vary continuously. Nevertheless the existence of attractor states leads to a well-defined problem and to a well-defined distinction between input patterns. As my studies show, the mechanisms described above by my equations can be realized, at least in principle, by suitably constructed lasers, chemical reaction networks, semiconductor devices and those of molecular microelectronics. The proof of the feasibility of such devices employs the concepts and methods of synergetics developed for the theory of spontaneous pattern formation in nonequilibrium phase transitions. I did not make any attempt here to compare my approach with those on explicit verifications of neural computers, nor did I make an attempt to enter the still controversial field in how far algorithms on serial-computers are still superior to network-like or neural computers. Lack of space just does not allow me to do this here but I hope to come back to these questions on other occasions.

References

With respect to spin glasses I refer the reader to the articles of this book, where the relevant further references can be found.

Part II

Natural Computational Systems

The Representation of Space-Time in the Human Visual System

J.J. Koenderink[1] *and A.J. van Doorn*[2]

[1]Department of Engineering Science, University of Oxford, Parks Road, Oxford OX1 3PJ, United Kingdom
[2]Physics Laboratory, Utrecht University, Princetonplein 5, PO Box 80.000, NL-3508 TA Utrecht, The Netherlands

1. Introduction

The subject of the representation of space and time "in the head" is extremely broad and elusive. In this short paper we - quite arbitrarily - cut down on the complications through the artifice of selective disregard of many very important aspects of the problem.

Basically, we only discuss the rather restricted and artificial case of the "visual field" of one eye under conditions of strict fixation. Moreover, we only consider the discriminative capabilities for very simple, abstract tasks of a geometrical nature. Thus we disregard qualities (e.g color) and cognitive factors. Then the visual field is a spatially two-fold extended and temporally sequential space of a rather nondescript character.

It will hardly be necessary to remind the reader of the fact that the "visual world" partakes of other modalities than the visual alone, moreover that it is three-dimensional, and full of qualities and meanings. However, one has to sacrifice something if one desires to obtain some scientific insight in the matter : even the visual field is a hard nut to crack!

2. Summary of empirical Evidence

We will consider three very different kinds of empirical evidence here : *anatomical data* that describe the visual system in terms of intricate spatially articulated material systems, *physiological data* for which we will focus on the visual system as a "signal processor" (although one strictly speaking does not know what the signals are, nor what is being processed - but the language of signal processing certainly helps a lot in expressing the evidence; the language should not be confused with the thing, however), and finally *psychophysical evidence.* Psychophysics is the science of the structure of behaviour in so far as this is directly influenced by the interaction of the visual agent with the physical world. Thus introspection plays no role in psychophysics as a science, it is at best a heuristic device.

2.1. Anatomy and Physiology

From the anatomy we know the structure of the eye as a basically camera-like device. It is not unlike a fully automated CCD camera although more functions are automated than is typical on the technical artifacts. Moreover, we know that the retinal irradiance distribution is sampled by a dense array of mutually autonomous photosensitive devices, the receptor mosaic. Thus at this very front-end part of the visual system we may be assured that different optical inputs will result in different simultaneous orders of electrochemical excitations of a very large cell population. The optic nerves contain a few million fibers each, much less than the number of photoreceptors. Thus many receptor signals must be considered as able to influence a single nerve fiber. A form of "multiplexing" appears to be unlikely from physiological considerations, thus the convergence and consequent loss of structure must be expected to be appreciable. From electrophysiology we know the type of convergence approximately. A single nerve fiber reacts to an extended but roughly localized portion of the retinal irradiance distribution, its so-called "receptive field" (RF). Coarsely speaking the RF signal can be understood as an "AC-coupled" finite element approximation to the Laplacian of a nonlinear function (a kind of "automatic volume control") of the irradiance distribution.

From electrophysiology we know that the optic nerve fibers connect to a veritable plethora of somatotopic maps. In each map the electrochemical activity at a certain position correlates most strongly with a specific patch on the retina in such a way that the topology is approximately conserved (hence the term "maps"). The maps tend to differ in characteristic ways : they seem to cater for different aspects of the spatiotemporal irradiance structure. Even within the maps we find a certain locally regular segregation of submodalities. *(E.g.* a map may be especially sensitive for movement, whereas the different directions of movement segregate on the local scale.)

It is hard to say what actually goes on in these areas, but one may guess that - apart from the obvious segregation - local and global operations of a correlational and/or grouping type must be important.

2.2. Psychophysics

Psychophysics is absolutely necessary in order to be able to interpret the functional significance of the anatomical and physiological data. In fact, most of our insights into the spatiotemporal structure are predominantly of a psychophysical origin.

Historically, *visual acuity* has been the most extensively studied visual faculty. It yields direct information concerning the functional grain size of the visual field. Near the optical axis the grain size is of the order of a minute of visual angle, at somewhat larger distances from the axis (the distance from the axis is generally known as the "eccentricity") the grain size proves to be close to a constant fraction of the eccentricity (about one hundredth of the eccentricity).

Since the early fifties the visual acuity has been extensively explored via the contrast sensitivity for spatial sine wave gratings. This method has the advantage that one obtains a numerical evaluation of the sensitivity as a function of a scale parameter particular to the stimulus, under the experimenter's control.

Similar methods have been exploited for temporal resolution. Here one meets with a curious phenomenon : in some cases the temporal grain depends very strongly on the operationalization. For instance, you may observe "flicker" up to 60 Hz or more, but although you can discriminate the flicker from a steady presentation, you cannot distinguish *temporal order* with anything near this resolution. Temporal order is already lost at rates that are an order of magnitude lower! In certain pathological cases (forms of *amblyopia* or "lazy eye") the same phenomenon occurs in the spatial domain : although the patient can resolve the detail he cannot make use of it because "the visual field appears scrambled" even though we can find no obvious faults with either the optics or the signal processing parts.

In psychophysics you can make fairly precise measurements of the detectability of spatiotemporal contrast as a function of outer (size, duration) and inner (spatiotemporal grain of the stimulus) scale parameters over extensive parameter ranges. (Say 10 decades of average retinal irradiance, several decades of eccentricity, idem of scale parameters). Such measurements are extremely time consuming and have not often been carried out on the necessary grand scale. (KOENDERINK et al. [1].) They have yielded a few very valuable insights, however. The results are most concisely expressed in terms of certain invariant properties : when you apply a spatial scaling and shift in the visual field, but leave all measures invariant as expressed in the retinal grain size (*i.e.* about one hundredth of the eccentricity) , then *all thresholds remain invariant*. (Notice that this requires a scaling of the irradiance too, because irradiance is radiant energy per area, *etc.*)

The capabilities of the human observer to deal with well-defined simple geometrical tasks of a topological, projective, affine, or metrical nature have hardly been studied in any systematic way. This is a real pity since one certainly needs such knowledge for further theoretical explorations.

A faculty that has attracted a lot of attention is the sensitivity to *movement*. One does not see the hour hand of a clock move, although one certainly notices a displacement after a finite time. On the other hand, one may perceive movement if an object assumes different positions at different times even if there is no evidence that the object ever assumed the intermediate positions! (This is of course the basis for the movie industry.) It is even possible to perceive movement without change in position (the so called "waterfall effect" is a well-known example). All such observations suggest that "movement" is elicited if certain "movement channels" are activated, - no matter how. We do have a rather precise notion about the structure of such channels nowadays. (VAN DOORN et al. [2].) Basically a motion-sensitive device in the human visual system is a correlator of spatially offset RF-signals that have suffered different amounts of delay in some way or other.

3. Theory

There does not exist a single, generally accepted theory of spatiotemporal representation in the head. It is even true that one can hardly observe a convergence of results or insights. The field is very much in flux and the value of many important notions that date from the previous century have not even been completely assessed. It is certainly true, however, that the modern electrophysiological methods yield data that allow novel tests of theories and that the pace at which the field is developing is certainly accelerating enormously.

3.1. The Basic Sampling System

The basic sampling system of the retina is composed of roughly circular RF's with a center/surround-oriented internal structure. These fields overlap extensively and are inhomogeneously distributed. Psychophysics allows an estimate of the RF density as a function of the eccentricity (e) and the RF diameter (s). (KOENDERINK et al. [3].) In a retinal area dA one has a number dN of RF's with diameters in the range [s,ds] such that :

$$dN = u(e,s)\, ds\, dA \quad \text{with}$$

$$u(e,s) = C/s^3 \quad \text{if } s \text{ exceeds } e/100 \text{ or } 1'$$

$$= 0 \quad \text{otherwise.} \qquad (1)$$

Notice that this is close to a self-similar density function : hence the psychophysical invariances mentioned above.

From the threshold data one cannot obtain more than a density function because the relevant theories treat the RF's as mere marbles in a bag without any spatial interrelations or order. However, the given density function admits of a very plausible interpretation : it is as if one has the convenience of possessing a multitude of retinae of uniform grain, some fine grained some coarse grained, but all of them containing the same number of sampling apertures (roughly 256^2 of them). The grain sizes would be uniformly distributed on a logarithmic scale (intervals of 10-20%). All "retinae" are centered on the optical axis. Such a system yields effectively a series of "pictures" of the same resolution (in terms of the number of pixels) but varying scope, from extreme wide-angle to the extreme tele-view. Thus the sampling system works as a "zoom-lens" in such a way that all zoom factors are simultaneously available.

3.2. The Simultaneous/Successive Order (Topology)

A technical CCD-camera has an inherently two - dimensional and ordered photodiode array. One wonders whether such a notion carries over to the optic nerve. Does every nerve fiber carry a unique label that specifies its place of origin? (The question was explicitly framed by LOTZE [4] in the 19th century and is generally known as the problem of "local sign".) We don't know, but it is possible to show that such would not be a necessary assumption. (KOENDERINK [5].)

As said above the RF's *overlap* considerably. Such an overlap must lead to a correlation of signals carried by nerve fibers that arise from similar locations and such a correlation can be measured by the electrophysiologist but is also an objective datum for the nervous system itself (unlike the "labels"). If the nervous system were to interpret correlation of signals as spatial overlap in some abstract space, the correlation structure could be used to compute a geometrical order in that space. Indeed many geometrical properties can be formulated as logical relations on the correlation structure. (For instance the relation of inclusion : "A is included in B" can be defined in the following way : A is included in B if and only if it is true that for every C such that C and A correlate it is also true that C and B correlate.)

It is possible to show that subject to certain mild conditions (basically the existence of sufficiently many RF's and a convex shape for the individual RF's) the correlation structure suffices as a basis on which to construct a topology, thus to find the homology, the dimension, *etc.* A well-known example is HUNTINGTON's [6] model of Euclidean geometry as a logical calculus over the totality of all circular discs (or balls in higher dimensions than two). The two-dimensionality of the visual field and its topological structure are thus objective properties of the optic nerve signals.

3.3. Scale Space

What can you possibly mean when you observe that oak trees have a hemispherical, cypresses a conical shape? Clearly this is nonsense because the shapes are extremely articulated arborizations instead of simple geometrical objects. The point is that the notion of shape depends critically on the resolution. If you sample a tree on the cm scale it is highly arborized, at the m scale it may well be hemispherical, at the km scale it is but a point.

This notion of the importance of scale fits in very nicely with the "zoom lens"-like sampling structure described above : the eye samples the irradiance distribution with various resolutions and sets up a set of charts like in a geographical atlas. The "world map" has a large scope but also a coarse grain, the other maps have a proportionally lesser scope as you increase the resolution. You may think of the retinal irradiance distribution as a function of the spatial coordinates (x,y) and a scale parameter (t) : we will write (x,y;t) in the sequel. This "scale space can be given a structure in natural ways. For instance one can define a natural mapping between sections of different resolution, just as one can point out Paris (say) on a map of Europe and a map of France. We call this "the natural correspondence".

It is clear that the natural correspondence cannot be one-to-one : the Quartier Latin and Montmartre on the plan of Paris clearly map on the same point of the world map! Notice that this cannot happen the other way around (e.g. London and Paris on the map of Europe cannot map to the same point of a higher resolution, regional map). Such a structure in which the natural correspondence only bifurcates towards the direction of increasing resolution has the structure of a "tree" in the parlance of computer science. We call such a continuous tree structure "causal".

We naturally require that scale space should be causal and this is a sufficient condition to prove that its structure is governed by the so-called *diffusion equation* (KOENDERINK [7]) , i.e. that the sampled irradiance I(x,y;t) is a solution of the partial differential equation

$$I_{xx} + I_{yy} = I_t \qquad (2)$$

(we use the notation $dI/dx=I_x$ etc.) with $I(x,y;0)$ (i.e. the stimulus pattern) as a boundary condition. Then the natural correspondence is defined through the integral curves of the vector field

$$\underline{s} = -I_x I_t \underline{e}_x - I_y I_t \underline{e}_y + (I_x^2 + I_y^2)\underline{e}_t \qquad (3)$$

where $(\underline{e}_x, \underline{e}_y, \underline{e}_t)$ denote the unit frame vectors for scale (that is (x,y,;t)) space. The correspondence is defined at all places except those where the irradiance gradient vanishes ($I_x^2+I_y^2=0$), *i.e.* the critical points of the irradiance. At these points you have bifurcations, as when in the case of the geographical atlas the conventional symbol for "city" that is used to mark Paris on the world map suddenly explodes into structure (Quartier Latin, Montmartre, ...) when you flip pages to the Paris town plan. The diffusion equation ensures that the branches are thrown off in the direction of increasing resolution.

To run the diffusion equation on an irradiance distribution is equivalent to a convolution of that distribution with a Gaussian kernel G_t. The difference between two such blurred irradiance distributions at slightly different resolutions is proportional to a convolution with $dG_t/dt=(d^2/dx^2+d^2/dy^2)G_t$, *i.e.* a "sombrero"-like profile of weights that closely resembles the structure of the retinal RF's. Thus the optic nerve appears to carry signals that can be interpreted as the derivative of scale space with respect to the resolution parameter. This makes obvious sense because it is effectively an economical encoding. Notice that when you integrate the optic nerve signal for all resolutions you obtain the original irradiance pattern as sampled at the highest resolution except for an arbitrary "minimal irradiance distribution", *i.e.* a distribution that satisfies the Laplace equation $I_{xx}+I_{yy}=0$.

3.3.1. The differential Structure

Consider how the retinal irradiance pattern could be processed in order to find significant structural properties. Local properties must be present in the differential structure defined on the tangent space of a point (KOENDERINK [8]), thus we come to consider the k-jet $j_k(I(x,y;t))$ at the point (x,y) for a given resolution t, i.e. the equivalence class of all irradiances I^* that agree with I in all partial derivatives up to the kth order at the point (x,y) for the resolution t. As a representative you may take the truncated Taylor series that locally approximates I, except for the constant term, that is

$$I(x,y;t) \sim I_x(x,y;t)\,x + I_y(x,y;t)\,y + 1/2*(I_{xx}(x,y;t)\,x^2 + ..)$$
$$+ .. + 1/k!*(.. + I_{y..y}(x,y;t)\,y^k)\,. \qquad (4)$$

It is clear that the numbers $(I_x, \; .. \;, I_{y..y})$ describe the jet completely, thus one way to represent the jet is to compute all the mixed partial derivatives up to the kth order.

The irradiance $I(x,y;t)$ is merely a blurred version of the stimulus $I(x,y;0)$, namely $I(x,y;t)=I(x,y;0)*G_t$. The mixed partial derivatives of the blurred function are

$$d^{n+m}/dx^n dy^m \quad I(x,y;t) \; = \; D^{n+m} I_t \; = \; D^{n+m}(I_0 * G_t) =$$

$$= I_0 * D^{n+m} G_t \; = \; I_0 * G_t^{nm} \qquad (5)$$

(Please take note of the fact that we use the convenient abbreviation "I_t", "G_t", *etc.* to indicate function at the level of resolution t. The subscript t does not indicate a partial derivative with respect to t here !) where the kernel G_t^{nm} is a product $G_t^{nm}(x,y)=g_t^n(x)g_t^m(y)$ of the one-dimensional kernels

$$g_t^n(u) = (-1/2\sqrt{t})^n H_n(u/2\sqrt{t})\exp(-u^2/4t)/\sqrt{4\pi t} \; . \qquad (6)$$

(The $H_n(u)$ denote the Hermite polynomials.)

The "fuzzy derivatives" G_t^{nm} behave in many respects like normal derivatives (notice that the operation G_t^{nm} is just $d^{n+m}/dx^n dy^m$ in the limit for $t \rightarrow 0$). In fact when you convolve the original function with such a fuzzy derivative you obtain the *exact derivative* of the *blurred function.* However, the fuzzy derivatives can be implemented with a finite RF-like weighting function and in a way "lift" the notion of derivative out of the realm of the infinitesimal. Some useful relations are :

$$d/dx \; g_t^n(x) = g_t^{n+1}(x) \qquad (7)$$

$$d/dt \; g_t^n(x) = g_t^{n+2}(x) \qquad (8)$$

$$g_t^n(x) * g_s^m(x) = g_{t+s}^{n+m}(x) \qquad (9)$$

$$I(x).g_0(x) = \sum_{k=0} \{ g_t^k * I(x,0)|_{x=0} \} 2^{k-1} t^{k-1/2} g_t^k(x)/k! \qquad (10)$$

$$F[\, g_t^n(x) \,] = (iw)^n \exp(-w^2 t) \; . \qquad (11)$$

Equations 7,8 express simple order raising relations that immediately reveal the relation to the diffusion equation (differentiating twice with respect to x has the same effect as differentiating once with respect to t). Equation 9 shows how the fuzzy derivatives behave under concatenation. Equation 10 shows how to reconstruct the original irradiance pattern from the blurred version within a small neighborhood defined through the window function $g_0(x)$, *i.e.* it solves explicitly the problem of running the diffusion equation backwards. Equation 11 expresses the fact that in the frequency domain the fuzzy derivatives appear as bandlimited differentiators (a true differentiator would have the spectral representation $(iw)^n$). In fact the spectral representation is often very useful. The fuzzy derivatives are very similar to sine (odd order) and cosine (even order) functions of circular frequency $w_n = \sqrt{[(n+1)/2]/t}$, modulated with a Gaussian envelope of halfwidth $4/\sqrt{t}$. The bandwidth is asymptotically equal to $1/2\sqrt{t}$ and the bandwidth-spatial width product is approximately unity : the fuzzy derivatives are indeed very similar to the well known "Gabor functions".

The exciting possibility is of course that it is feasible that the cortical RF's do represent a jet-extension of the scale space based on the retinal irradiance distribution. The partial derivatives at a point are represented through ensembles of receptive fields having profiles of the type G_t^{nm} which do in fact very much resemble the RF profiles that have actually been determined empirically. It would appear that all derivatives up to at least the third, perhaps the fourth order are represented. Such a local ensemble can be regarded as a local approximation of the irradiance as sampled at a given grain size. If you are interested in certain local shape descriptors you may write down their structure in terms of the local differential geometry and then substitute the activity of RF's for the partial derivatives in order to obtain the correct neighborhood operator. This is an extremely powerful device since all local shape descriptors can be expressed as algebraic combinations of mixed partial derivatives of the irradiance. Thus the differential geometry can be directly translated into neural network implementations.

Consider an example : "boundary curvature". From elementary differential geometry you find that the curvature of an equi-irradiance contour is

$$k = - \frac{I_y^2 I_{xx} - 2 I_x I_y I_{xy} + I_x^2 I_{yy}}{(I_x^2 + I_y^2)^{3/2}} . \qquad (12)$$

This expression simplifies appreciably in suitable coordinates, *e.g.* you may rotate the coordinate frame such that I_x vanishes. Then the curvature is just the ratio of I_{xx} and I_y. Thus the recipe for boundary extraction in the cortex is :

ALGORITHM I (*boundary curvature*) :
1/ Find a coordinate system such that $I_x=0$;
2/ Take the ordered pair (I_{xx}, I_y) as a measure of the curvature.

Notice that this procedure entails only lookups, no computation whatsoever! The procedure might very well be carried out in the cortex in this way because the cortical representation of the jet is in exactly the right form to sustain it (*vide infra*).

The cortical representation does not appear to make much use of preferred Cartesian coordinate systems. Instead, the first order derivatives are represented through a set of directional derivatives I_u for a great many directions $\underline{e}_u$. Similarly the second order derivatives are represented via a set of second order directional derivatives I_{uu}. (Thus the mixed derivative I_{xy} is not represented at all! It can easily be obtained as a linear combination of the type $I_{uu}-I_{vv}$, however.) This type of representation is not necessarily economical hardware-wise, but it is extremely suitable for algorithms that require an adapted coordinate frame (*i.e.* almost all algorithms!). It has several other advantages having to do with the robustness (noise sensitivity, resistance against component failure, ...) that we will not discuss here.

A final remark regarding these jet extensions has to do with the meaning of individual RF signals. In the physiological literature one often encounters the notion of "feature detectors", *e.g.* the RF's of type G_t^{20} are generally known as "line detectors" or "orientation detectors". This appears to be an overly limited and indeed misleading view : the same RF (or mixed partial derivative) can subserve many different functions according to the visual routine of present interest - it does not have a fixed "meaning". (For instance the boundary curvature scheme discussed above does not use G_t^{20} as a "line"- or "orientation"-detector at all!) The meaning of a single RF is not fixed but depends on the activity of its mates and the goal to which it will be used. The "meaning" is very much in the use.

3.3.2. Connexions (Multilocal Structure)

In the schemes discussed in the previous paragraph the shape descriptors were taken to be algebraic combinations of mixed partial derivatives at a point, *i.e.* they were considered as essentially point properties. In many cases one needs descriptors that do have an essential *multilocal* character. Consider for instance the boundary curvature again. Suppose one had merely a first order jet extension, then there would be no way to implement a boundary curvature routine on purely local data. On the other hand a bilocal scheme would do the job quite simply, the following scheme could be used :

ALGORITHM II *(boundary curvature)* :
1/ Find the direction $\underline{D}_u$ for which $I_u=0$;
2/ Take a step du in the direction u and find the direction v for which again $I_v=0$;
3/ The ratio of the orientation difference between v and u with the stepsize du equals the boundary curvature.

This works just fine, but the differences with the previous scheme are worth considering. Several assumptions have been made :

- it is assumed that you know how to perform a step in the direction u in the first place, i.e. how to identify an adjacent jet extension in a given direction;
- it has been assumed that you can compare orientations at different positions and that you can assign a value to an orientation difference.

These assumptions may appear comparatively trivial, but in fact no one has the slightest notion how such abilities could be implemented in the neural system. Clearly their representation has to take the form of a system of pointers (in the abstract sense, we do not suggest a possible implementation here!) between RF's, *e.g.* pointers between RF's of similar orientations at different positions.

In the mathematical (differential geometrical) sense one needs a *connexion,* if not the classical connexion (the complete set of Christoffel symbols) then at the very least a projective structure.

Similar considerations apply to the temporal domain, although *e.g.* the notion of "simultaneity" needs extra consideration. Simultaneous events can only be appreciated at a point, and events at different places cannot be compared at these places themselves, in fact the finite transmission delays effectively decouple different places if you regard their signals at identical times. Simultaneous events at two places can be appreciated (after the fact!) at a third place that receives signals with equal transmission delays from the two former places. We know very little about such structures in the nervous system, apart from the fact that the motion-sensitive system appears to be based on the comparison of signals that arrive at a given place from two (or more) other places with different delays. The former place detects effectively coincidences *(i.e.* simultaneous events) and is in itself in no position to appreciate the fact that "simultaneity" really arises because of different moments of stimulation at different places!

3.4. The Global Structure

We have now discussed what may be termed "punctal" (i.e. residing in the tangent space of a point) and "local" (i.e. depending on a connexion , or system of pointers to adjacent jets) properties. There do in fact exist many phenomena that have been described in the psychology of vision that have to be termed "global". Such phenomena include the comparison of features at very disparate positions in the visual field or the judgment of straightness of very extended one-dimensional entities, *etc*. Very little is known about such phenomena. Observers are able to make judgments about straightness, but it is well known that phenomenally straight lines are in no way great circles in the visual field, in fact they appear to be rather well described by the system of small circles that have the antipodal point of the visual axis in common ("Occipitalpunkt"). This has been related to the vagaries of the human eye-movement system by VON HELMHOLTZ [9] and this has remained the only sensible "explanation" up to date. It is simply not known whether these phenomenally straight lines are equal to the system of "autoparallels" that can be obtained from the connexion through integration or not.

Other examples include the judgment of "co-circularity" of orientations at widely different places. Observers can do this quite well, in fact they seem to base their intuitive estimates of the "medial axes" of shapes on this ability. (Co-circularity obviously defines a local axis of mirror symmetry.) Such an ability suggests that observers are able to perform a parallel transport of their tangent spaces over arbitrarily large stretches. However, the holonomy group defined by this parallel transport has never been experimentally explored.

Many other capabilities that reflect global processes have to do with the "Gestalt laws" from experimental psychology. An example would be the "grouping" of spatially disparate shapes on the basis of some geometrical similarity. A critical analysis of these phenomena in terms of geometrical abilities of different kinds has never been made, despite the obvious importance.

4. Conclusion

In conclusion it seems fair to say that although we have gained some insight into the nature of the representation of space and time "in the head", there remains much work to be done. Some clarity has been obtained regarding the local topological and differential properties, but even the subject of the nature of the connexions remains very much in the dark. Regarding more global problems our ignorance is profound. These problems are both of a theoretical and of an empirical kind. Obviously the theorist needs empirical evidence to guide his steps. However, it is historically evident that the problems in question are so intricate that it appears next to impossible to devise significant experimental paradigms from purely phenomenological reasoning instead of employing some guidance from the theoretical side. In this case it seems prudent to let the theory develop a bit ahead of the experiment in order to generate useful alternatives instead of pursuing empirical goals on a completely *ad hoc* basis. A special problem in this field is that the conventional walls between the faculties largely prevent a fruitful dialogue between the theorist (mathematicians, physicists, ..) and experimentalists (typically psychologists, physiologists, ..) because of their very different training. Future progress will heavily depend on a much more intensive interaction between these different fields.

5. Literature

1. J.J. Koenderink, M.A. Bouman, A.E. Bueno de Mesquita and S. Slappendel : J.Opt.Soc.Am. 68, 845-865 (1978)
2. A.J. van Doorn and J.J. Koenderink : Exp.Brain Res. 45, 179 - 195 (1982)
3. J.J. Koenderink and A.J. van Doorn :Biol. Cybernetics 30, 157-167 (1978)
4. H. Lotze : Mikrokosmos (Hirzel, Leipzig 1884)
5. J.J. Koenderink : Biol.Cybernetics 50, 35-50 (1984)
6. E.V. Huntington : Math.Ann. 73, 522-559 (1913)
7. J.J. Koenderink : Biol. Cybernetics 50, 363-370 (1984)
8. J.J. Koenderink : Biol. Cybernetics 55, 367-375 (1987)
9. H. von Helmholtz : Handbuch der Physiologischen Optik (Voss, Leipzig 1896)

Neuronal Mechanisms of the First, Second, and Third Order Contrast in the Visual System

G. Baumgartner, E. Peterhans, and R. von der Heydt

Department of Neurology, University Hospital,
CH-8091 Zürich, Switzerland

1. Introduction

Illusions have been used extensively in sensory physiology as an access to the functional organization of sensory systems. This is shown for example by MACH's bands and his explanation of this phenomenon by a negative lateral retinal interaction. They are useful also in neurophysiology and many can be correlated with the activities of single neurons at different levels of processing (BAUMGARTNER /1/; v.d.HEYDT et al./2/; PETERHANS et al./3/).

During the last thirty five years much effort has been devoted to studies of the transfer of visual information from the retina to and within the hemispheres and the processing involved (JUNG et al./4/; HUBEL and WIESEL /5,6/; JUNG /7/; ZEKI /8/). Since the sensory input to the retina via 10^7 receptors is much more than its output via 10^6 ganglion cells, one had to expect that the retina does not simply transform light energy into neuronal signals, but rather being a miniaturized version of the brain acts as a processor with high capabilities. To avoid substantial loss of information processing has to start immediately at the receptor level. The question was then, where within the visual system neuronal correlates of different sensory phenomena originate or can be found. In the following we shall describe neuronal mechanisms which we believe to be important for the generation of illusions of achromatic border, areal and illusory contrast.

2. Neuronal Correlates of Simultaneous Contrast

1953 KUFFLER /9/ described two versions of receptive field organizations in the cat's retina. Receptive fields are defined as the area of the retinal surface from which the activity of a cell can be modified. He found that light-activated neurons were excited from the center and inhibited from a nearly circular surround (on-center-neurons), while light-inhibited neurons were inhibited from the center and activated from the surround (off-center-neurons). The effect of the center being always dominant and therefore responsible for the reaction to diffuse illumination. This meant a physiological support of MACH's inhibitory retinal interactions but left open the contribution of off-neurons in this context. We began therefore to study single neurons in the optic nerve, the thalamic relay of the visual pathway, i.e. the lateral geniculate body, and in the first visual area in the cat under conditions of simultaneous contrast. In any level concentric receptive fields of the kind KUFFLER /9/ described were found. It turned out that off-neurons were always activated on the dark and on-neurons on the bright side of border contrast (Fig.1). We proposed therefore that simultaneous contrast was processed first in the retina and transferred centrally via two neuronal channels. One, via on-center-neurons coding for brighter, the other, via off-center-neurons, for darker within its receptive field center. Since the surround activity in retinal ganglion cells was shown to be a function of adaptation and disappears in the dark adapted state (BARLOW et al./10/), we assumed that in the lateral geniculate body contrast was reproduced via inhibitory connections from retinal on-center cells from the field surround to on-

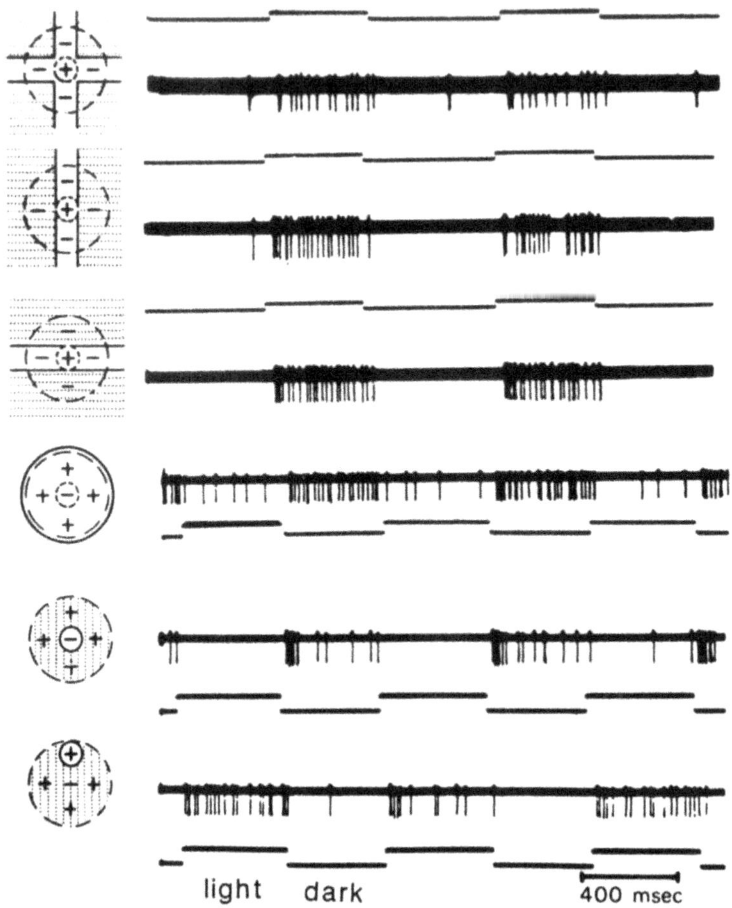

Fig. 1

On-center-neuron (upper 3 lines) and off-center neuron in different contrast illumination. The on-center neuron is shown by stimulation with a horizontal and vertical bright bar and a bright cross of equal luminance. Due to increased lateral inhibition the activity in the intersection is reduced.

Off-center neuron after stimulation of the whole receptive field and stimulation with a light spot in the center and in the periphery. Complete inhibition after central illumination, activation after illumination in the periphery. (Dashed circles = receptive fields, + = excitatory, - = inhibiting regions, continuous circles = light stimulus)

center cells and excitatory connections from retinal on-center cells from the surround to off-center neurons in the lateral geniculate body. HERMANN's illusion can easily be understood by this concept (Fig.2).

3. Neuronal Correspondencies of Areal Contrast

1959 HUBEL and WIESEL /5/ described the reorganization of receptive fields in the first cortical processing area (V1) along an axis of specific orientation. They found this orientation identical in all cells of a column orthogonal to the cortical surface. A given retinal area was represented by a set of columns of

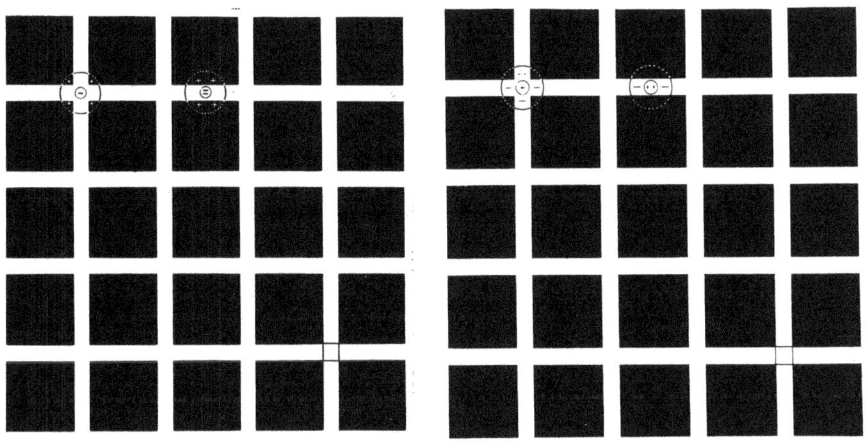

Fig. 2

Hermann grid: the darker intersection in the bright version is explained by the reduced activity of on-center neurons by increased lateral inhibition, the brighter intersections in the dark version by reduced activity of off-center-neurons by reduced lateral activation (see fig. 1)

different orientation. The existence of concentric fields in the cortex was first denied but finally agreed on. But the fact remained that only neurons in the input layer of IVc had concentric receptive fields. The majority of cells had oriented receptive fields with balanced excitatory and inhibitory input and did not respond to diffuse illumination. The processing of simultaneous border contrast had to undergo a transformation and seemed to be lost. It can be found again if one considers areal contrast (Fig.3)

Area contrast must be generated within the cortex. It is not found in the activity of afferent neurons. Neurons in the afferent pathway with receptive fields within a homogeneous area without light gradients are responding with the same activation irrespective of whether the area appears brighter or darker by contrast. Neurons of cortical cells with receptive fields within such an area do not respond at all, since excitatory and inhibitory inputs are balanced. Only the cortical neurons with oriented field corresponding to the orientation of the borders are activated under such conditions. We assumed therefore that areal contrast is generated via an extrapolation of the brightness across the contrast area from border to border depending on the activation of cells on the borders. The prerequisite for such an interpretation is that neurons coding either for brighter or darker are still separated and present. We assume therefore that cells with excitatory centers along the receptive field axis are further coding for brighter and cells with inhibitory centers along the axis for darker. Under this condition the direction of the luminance gradient at the border should define the apparent brightness of the field. Area contrast can so be understood as an augmented border contrast by generalization over space, or vice versa, border contrast as a special case of area contrast. Reversing the contrast gradients at both sides of a contrast field should lead to an apparent brightness in between, as it can be seen in Fig.3.

Two contrast fields of the same extension appear different in size, the darker smaller than the brighter one. This may be understood if one assumes that spatial localization is defined by the central axis of the receptive field. Neurons with excitatory field axis should indicate precisely the site of incoming light. Neurons with inhibitory axis, however, are only activated and therefore express their spatial code via lateral activation due to a light gradient from

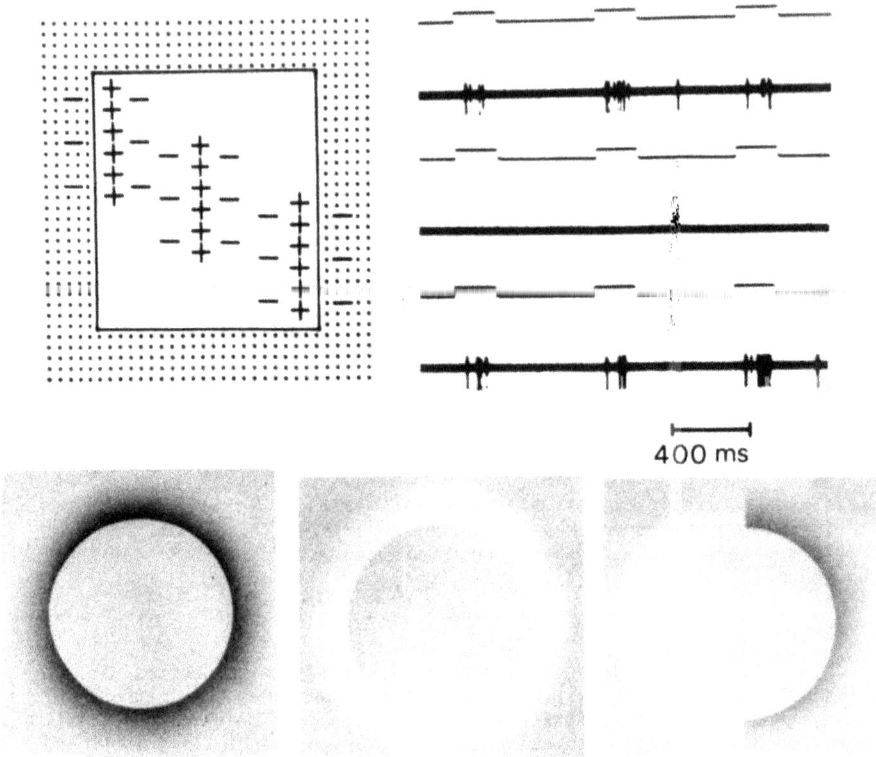

Fig. 3
Craik Cornsweet illusion of areal contrast, depending on the direction of contrast gradients at the border. The discs are of equal luminance. Inversed gradients at opposite sides weaken the illusion. Above, simple field neuron at the borders and within a bright contrast field. The line above the neuronal response at the right indicates the photocell. Line up = light on. The neuron responds only at the border (+ = excitatory, - = inhibitory areas of the receptive field)

the facilitatory border region to the inhibitory center area. This means that light is used to indicate darker in the less illuminated neighborhood. Lateral activation therefore must produce a systematic spatial shift toward the dark area or away from the bright one. This may explain the illusion.

From area contrast we can also conclude that the cortical processing discards redundant information over areas without brightness gradients. Therefore, a loss of nerve cells in V1 should not be realized in such conditions, since higher processing areas receive physiologically no information from the contrast area and should not be able to discriminate if this is due to loss of cells or only to non-activation of the responsible cell area. Clinically this corresponds to the experience that cortical scotomata, contrary to peripheral ones, are not noticed.

4. Neuronal Responses to Illusory Contours

The lack of afferent information due to a lesion within the afferent pathways is perceived immediately. When the destruction of cortical neurons remains unnoticed, the question arises, where this peculiar combination between the activity of neuronal cells and perception originates. How far has the neuronal processing to go before it reaches perceptive equivalence or are perceptive features not traceable at all on single neurons but rather are hidden behind the cooperative activity of large neuronal populations? Where is the activity of cells, which represent the borders of objects in discrete elements and which for a continuous border are spatially not continuously represented, as HUBEL and WIESEL /6,11/ have shown in V1, convergent or where does it become cooperative? In order to address this question further, we assumed that neuronal activity which is correlated with the perception of a bright bar should signal a bright bar as long as we see it no matter whether it is physically existent or an illusory construct. We therefore began to use stimuli which produced illusory contours and looked for neuronal correspondencies in the awake monkey, which can learn to discriminate them (Fig. 4). It was found that cells of V1 responding to a real bar did not respond to illusory ones (Fig.5). By contrast, approximately 30% of neurons in V2 did respond to the features which generated an illusory contour of the same localization and orientation as an activating real stimulus. The activation was in the average about 60% of the activity with real stimuli. If one manipulates the inducing stimuli such that the illusory contours disappear, the activity of the neuron disappears (Fig.6).

To relate the reaction of V2 neurons more precisely to the illusory contour and to separate it from the orientation of the features generating it, we studied these neurons by using an anomalous contour produced by abutting line gratings. Under these conditions a subjective contour appears orthogonal to the orientation of the inducing grating. It could be shown that the orientation of the subjective contour coincides with the orientation of a real contour which activates the neuron (Fig.7). Using oblique gratings did weaken but not abolish the reaction of the neuron as it weakens the illusion. The tuning curves for real and subjective contours were overlapping (Fig.8).

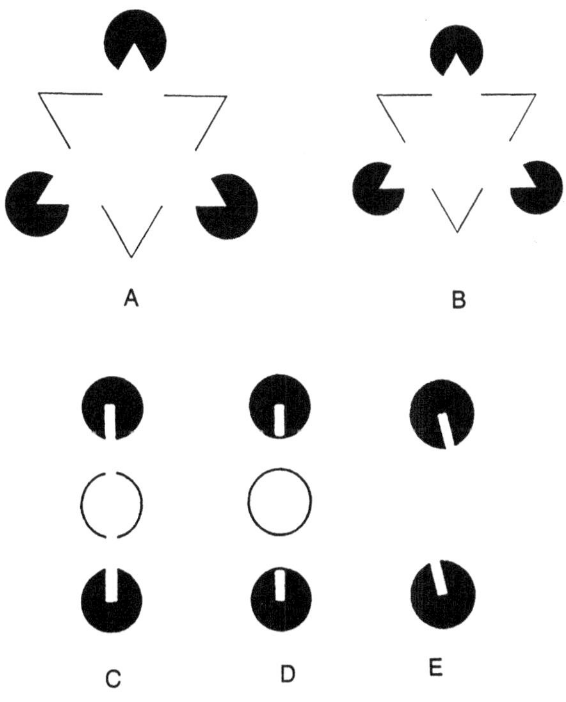

Fig. 4
Kanizsa (A, B) and Smith-Over illusions (C, D, E) and the disappearance of the latter by closing the circle or the gaps (B) or disarrangement of the gaps (E)

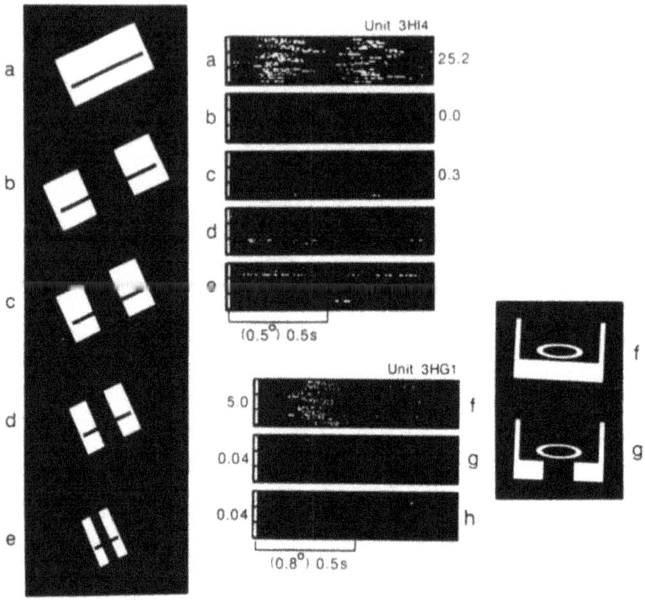

Fig. 5
Simple field neurons of V1 responding to a dark bar (Unit 3H14) or an edge (Unit 3HG1). No reaction to an illusory bar or edge. The dots indicate neuronal discharges and every line of dots corresponds to a movement of the real or apparent stimulus in both directions orthogonal to the orientation of the bar or edge across the receptive field. The few responses in d and e are due to small eye movements, the smallest gap in e being 20 min. of arc. The numbers give the average discharge rate per one stimulus cycle

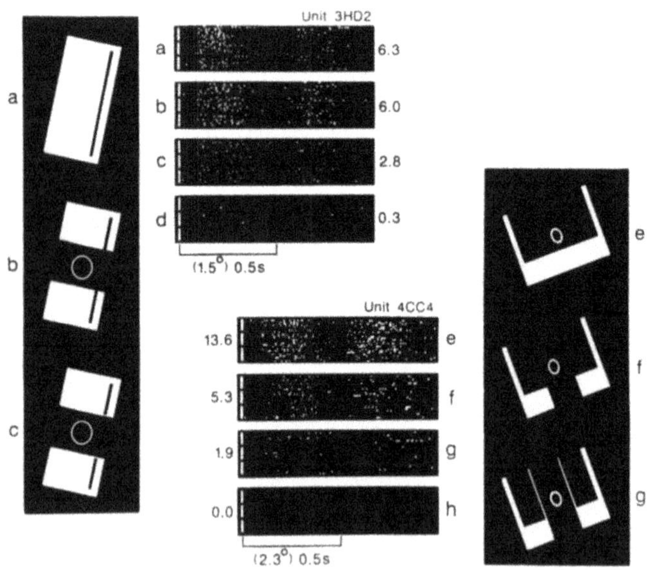

Fig. 6
Neurons of V2 responding to a dark bar and edge and the corresponding illusory contour. Note the strong reduction of activity by closing the gaps (c) or interrupting the illusory contour by thin white lines (g), d and h = spontaneous activity (see fig. 5)

An illusory contour disappears by small changes, e.g. by closing the inducing gaps. This shows that the neuronal response cannot be explained by length summations, since the change of luminance by closure is negligible. This is confirmed by the grating stimulus. Again the reaction of V2 neurons cannot be explained by anisotropic filtering or length summation since the brightness along the illusory lines is equal on both sides. When overlapping, separated and exactly abutting gratings were used, V1 cells responded depending on their reaction to bright or dark bars to overlapping or separated gratings, but did not indicate

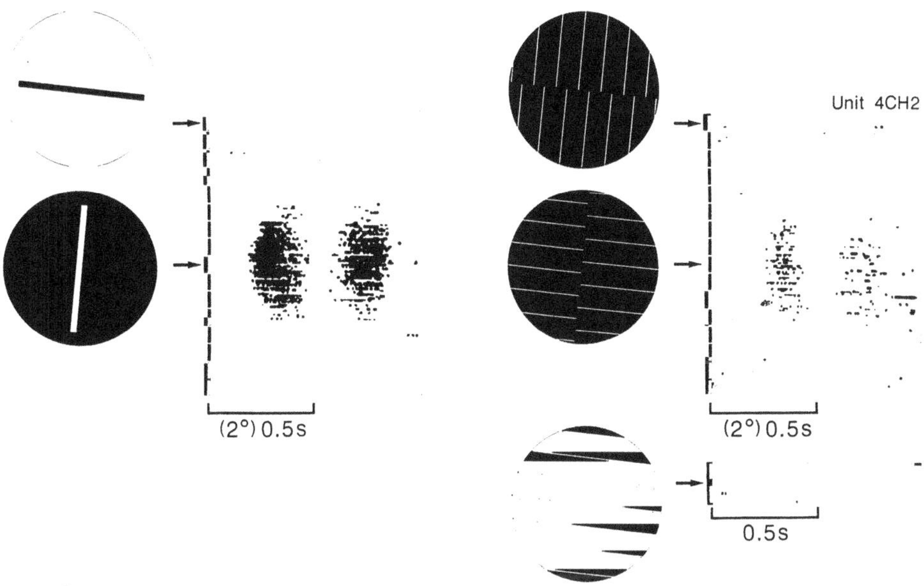

Fig. 7
V2 neuron responding to a real bright bar and to an illusory contour of the same orientation. The record beside the continuous lines shows the missing response if the lines are given alternating in the position of the abutting lines above

contacting ones. V2 cells, however, responded in addition also to the abutting gratings. This indicates in a very qualitative way that V2 cells take into account conventional receptive field mechanisms from V1 but in addition depend on more complex interactions. These cells are obviously influenced in a very defined way from areas outside their conventional receptive fields measured by real stimuli. They must be able to integrate by not yet understood algorithms the activity of an array of differently oriented and spatially distributed cell columns in order to do so.

5. Conclusion

From these data we conclude that simultaneous border contrast is prepared in the retina, reproduced in the lateral geniculate body and transferred to layer IVc of the visual cortex (V1) via two channels coding for brighter or darker. There this information is transformed to generate area contrast via some oriented cells still coding for brighter or darker. The visual scene seems to be treated in V1 by a huge amount of filters for detailed features, which discard redundant information and remind us of the ideas of D. MARR's /12/ computional approach. V1 seems to be still a preliminary processing area, as the missing reaction to illusory contours indicates. The neuronal mechanisms, which underlie illusory contours occurring in V2 indicate again an exclusion of redundant information. One could speculate that structures seen under these conditions are the most likely ones in a normal visual surround where occlusion by overlying objects with or without gradients of luminance or texture are abundant. The mechanisms generating apparent contours are important also for the perception of form of real objects where the shape remains unbiased although in some regions contrast gradients may be zero. In addition they may contribute to the stratification of depth (KANIZSA /13/) which is not lost after exclusion of disparity, motion parallax and other canonic depth cues. Since form is the primary feature in vision, it is conceivable that it is generated early in the cortical processing.

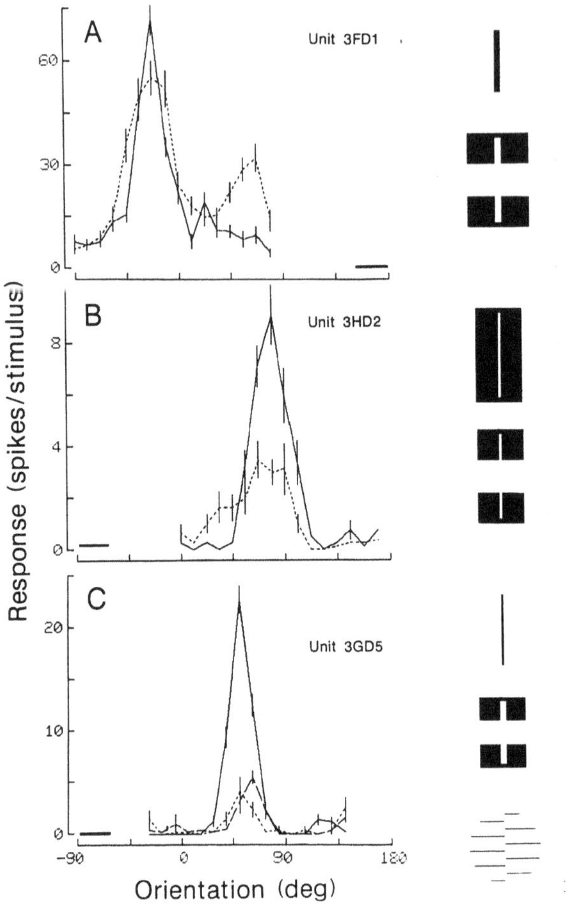

Fig. 8
3 neurons of V2 responding to real and illusory bars, their tuning curves for orientation. The neuron in C responded also to abutting gratings. The short horizontal bars indicate the level of spontaneous activity. (Solid lines = real bar, dotted lines = illusory contour stimulus, dashed lines = illusory contour with grid)

How it is integrated in the whole complex of the perceptional array and the necessary cooperation of neurons of distributed cortical areas, however, remains to be explored and explained.

6. Literature

1. G. Baumgartner: In Neurophysiologie und Psychophysik des visuellen Systems, ed. by R. Jung and H. Kornhuber (Springer, Berlin, Göttingen, Heidelberg 1961) p. 296-313
2. R. von der Heydt, E. Peterhans, G. Baumgartner: Science 224, 1260-1262 (1984)
3. E. Peterhans, R. von der Heydt, G. Baumgartner: In Visual Neurosciences, ed. by J.D. Pettigrew, K.J. Sanderson and W.R. Levick (Cambridge University Press, Cambridge 1986) p. 343-351
4. R. Jung, R. v. Baumgarten, G. Baumgartner: Arch. f. Psychiat. u. Ztschr. Neurol. 189, 521-539 (1952)
5. D.H. Hubel, T.N. Wiesel: J. Physiol. (London) 148, 574-591 (1959)

6. D.H. Hubel, T.N. Wiesel: Proc. Roy. Soc. B198, 1-59 (1977)

7. R. Jung: In Psychopathologie mimischer Gestaltungen, ed. by H.H. Wiesel (F.K. Schattauer, Stuttgart, New York 1974) p. 27-88

8. S. Zeki: Proc. Roy Soc. B217, 449-470 (1983)

9. S.W. Kuffler: J. Neurophysiol. 16, 37-68 (1953)

10. H.B. Barlow, R. Fitzhugh, G.W. Kuffler: J. Physiol. (London) 137, 338-354 (1957)

11. D.H. Hubel, T.N. Wiesel, M.D. Stryker: Nature 269, 328-330 (1977)

12. F. Marr: In Vision (W.H. Freeman, San Francisco 1982)

13. G. Kanizsa: In Organization in Vision (Praeger, New York 1979)

A Model of Figure/Ground Separation Based on Correlated Neural Activity in the Visual System

H.J. Reitböck, R. Eckhorn, and M. Pabst

Angewandte Physik und Biophysik, Philipps Universität, Renthof 7, D-3550 Marburg, Fed. Rep. of Germany

1 Introduction

Humans and animals can recognize objects even in viewing situations where the objects are seen against a patterned background and when several objects are adjoining or intersecting each other. An extreme example illustrating the figure/ground separation power of the human visual system is the well-known picture of a dog by R.C. JAMES (Fig.1). When one looks at the picture, it seems as if the regions belonging to the object "dog" were somehow labeled to the extent that even contiguous black or white areas appear to be divided into sections that belong to the object and into others that belong to the background.

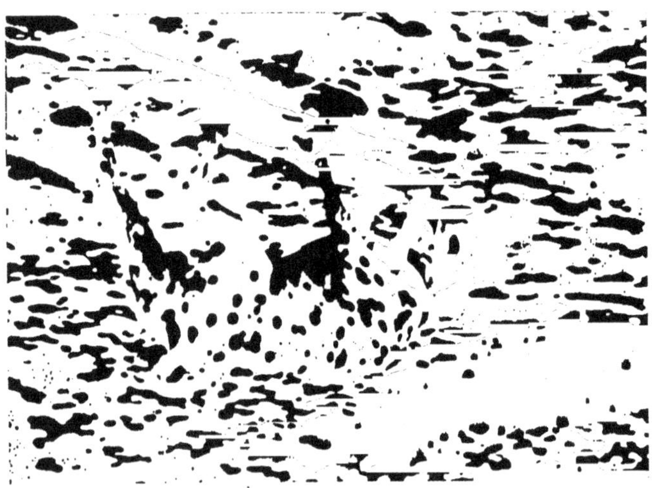

Fig.1 An extreme case of figure/ground similarity. (R.C. James, 1966)

Although figure/ground separation is performed in the visual system without any noticeable effort, it is a taxing problem in machine pattern recognition, because a curious situation of infinite regress appears: In order to recognize (classify) the object, a more general, invariant object description has to be generated that can be compared with data in memory. This, generally, requires that the object has been separated from the background. In order to separate it from the background, the object boundaries must be determined, i.e. a specific region in the scene has to be recognized already to represent the object.

But how is that region to be defined if the object is not yet known? The contradiction can be resolved, if *a priori* information is available about the object. In the simplest case, if an exact template of the object is available, cross-correlation can be used to determine whether the scene contains the object, and if so, to indicate its position. But this is pattern detection and not pattern recognition. Closer to pattern recognition is a size-invariant correlation as can be realized e.g. via the Mellin transform and via other transforms that are based on logarithmic space scaling [1-4, 40]. These methods, however, become impractical if the object is three dimensional and should be recognized from different viewing positions and if the object can assume a large variety of shapes. This, of course, is the usual pattern recognition situation the visual system is faced with.

We are proposing a model for figure/ground separation in the visual system that is based on the assumption, that correlated neural activity is used as a label for region definition. In Fig. 1, regions that belong to the object "dog" would be labeled such that correlated or even quasi-synchronous events would be induced in the temporal discharge patterns of neurons coding these regions. Although this assumption is speculative, it is consistent with basic neural properties (neurons can act as coincidence detectors) and it is well suited to model certain pattern recognition operations, such as the iterative formation of interpretation hypotheses using *a priori* information via feedback from various processing levels, and for the associative linking of object attributes.

In the following, two successive stages of the model will be discussed: a.) the generation of texture-specific temporal activities, and b.) a hierarchically organized network of correlators with feedback that transiently link domains of coherent activities, such as neural assemblies activated by regions of similar texture.

In Section 3.2 we will show that an elementary implication of Stage I of the model, already, can explain pre-attentive texture discrimination in agreement with psychophysical results. In Section 3.4 neurophysiological evidence for different degrees of correlated activity in the visual system, as assumed in the model, will be given. These data were obtained with multi-electrode recordings [37, 41- 43] in our laboratory [38, 39].

2. Correlation Theories of Brain Function

Since REICHARDT's pioneering model of the visual system of the fly [5], many models of neural correlation detectors have been developed. More global correlation theories of brain function have been proposed by v.d. MALSBURG [6, 54], JOHANNESMA et al. [7], van de GRIND et al. [8] and others. The concept of "local sign" [9], also, could be realized via local correlations. For the visual system of the fly, REICHARDT et al. [10-13] have developed a model of figure/ground separation based on relative movements between texture regions. A concept for region labeling via correlated neural activity has been proposed by REITBOECK [14, 56] and a model of pre-attentive texture discrimination based on this concept has been developed by REITBOECK et al. [15].

3. The Model

A moving texture region produces correlated or even quasi-synchronous activity in those visual neurons that are stimulated by it. Correlated activity could thus be a mechanism to "highlight" something, to make it stand out. Even if there is no relative movement between texture regions, the retinal image is always shifting relative to the retina due to eye and body movements. This results in temporal variations of the activity of the receptors and of subsequent visual neurons. Spatial texture intervals are thus converted into temporal intervals within spike trains. The discharge patterns of neurons that are activated by similar textures, therefore, should be similar.

Our model is based on the hypothesis that correlated activity is used in the central nervous system as a label for the associative linking of signals. The shape of the correlation function, particularly the width and magnitude of correlation peaks is assumed to be a measure for the specificity and magnitude of the associative link within and between the regions and modalities encoded by the signals. The model applies to early as well as to higher processing stages in the CNS:

3.1 Stage I: The Generation of Texture-Specific Temporal Signals and Their Separation by Coherent Temporal Sampling

At peripheral processing stages, temporal correlation is assumed to be stimulus-induced and associations are established between neurons that are activated by similar physical stimuli. At the retinal level, a moving texture will produce correlated activity that predominately depends on the texture intervals, on the speed of movement, on the type and overlap of receptive fields of neighboring neurons, and on the lateral interconnections. It is well known that retinal ganglion cells sensitively respond to stimulus motion. In the cat retina, ECKHORN and POEPEL [16] have shown that temporal modulations of the discharge of X cells are induced even by minute movement amplitudes in the range of the eye tremor, and in Y cells such modulations have been recorded up to the very fast saccadic velocities.

3.2 Computer Simulation of Stage I: (Pre-Attentive Texture Discrimination)

The performance of Stage I of the model in texture discrimination tasks has been tested with several dozens of pre-attentively discriminable and indiscriminable texture pairs, including those published by JULESZ [17-20]. Examples of region characterization via interval histograms and region separation are shown in Figs. 2-5.

In these examples, each field of 16x16 pixels contains one texture element, such as an M, W, L, etc. The textures are shifted relative to a "retina" (N_1 in Fig. 6) whose "neurons" respond to brightness transients. The total shift is 16 pixels. For each "neuron" the intervals between brightness transients are plotted in a histogram; higher order intervals at the single "neuron" level also contain concatenations of basic intervals. Figs. 2-5 show global interval histograms (sum of the histograms of individual "neurons") for a given scan direction.

The "neurons" in N_1 are assumed to have ON-response, producing a spike when a 0/1 transition moves over its "receptive field". The distances between 0/1 transitions in shift direction and the shift velocity, therefore, determine the spike intervals. Temporal sampling in a subsequent layer N_2 (e.g. via an array of AND gates with a common strobe input) yields region separation, if the sampling is done at an interval length that has different probability in the two texture regions. Region separation with this method is shown in Figs. 2e-5e. At the "single neuron level", longer intervals are also permitted to result from concatenations of intervals. N_2, therefore, can be realized as in Fig. 6 without additional circuitry to exclude higher order intervals. A more detailed texture characterization can be obtained via the global interval histograms for several scan directions. The

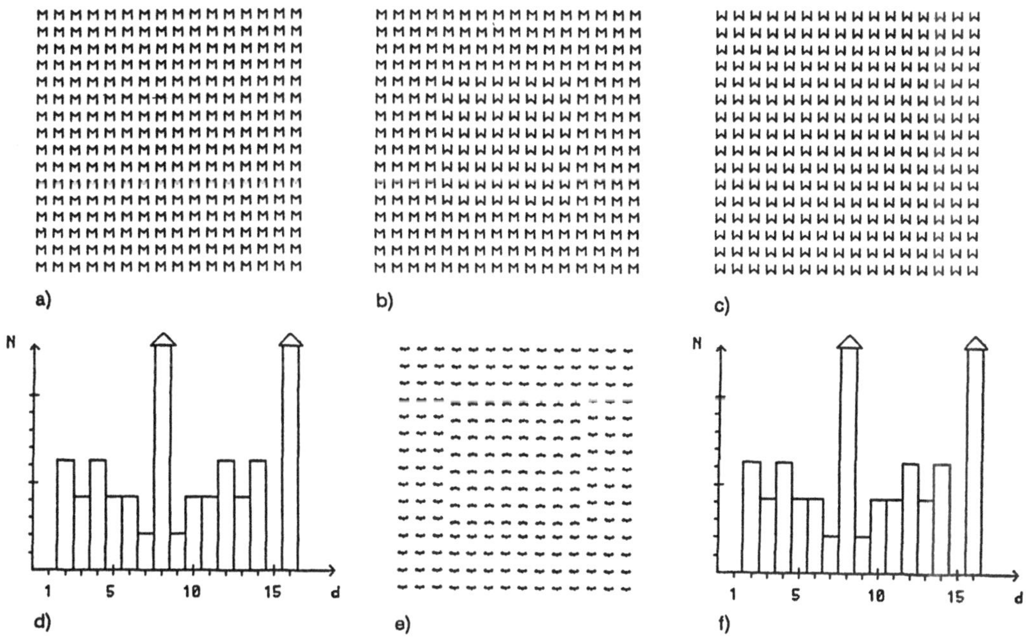

Fig. 2: a) Texture A. b) Texture pair A/B. c) Texture B. d), f) Interval histograms for textures A and B (horizontal scan direction). Intervals between black/white transitions in pixels. (↑ indicates: values are exceeding the display range.) e) Spatial distribution of horizontal intervals of length d = 2 pixels in the texture pair A/B.

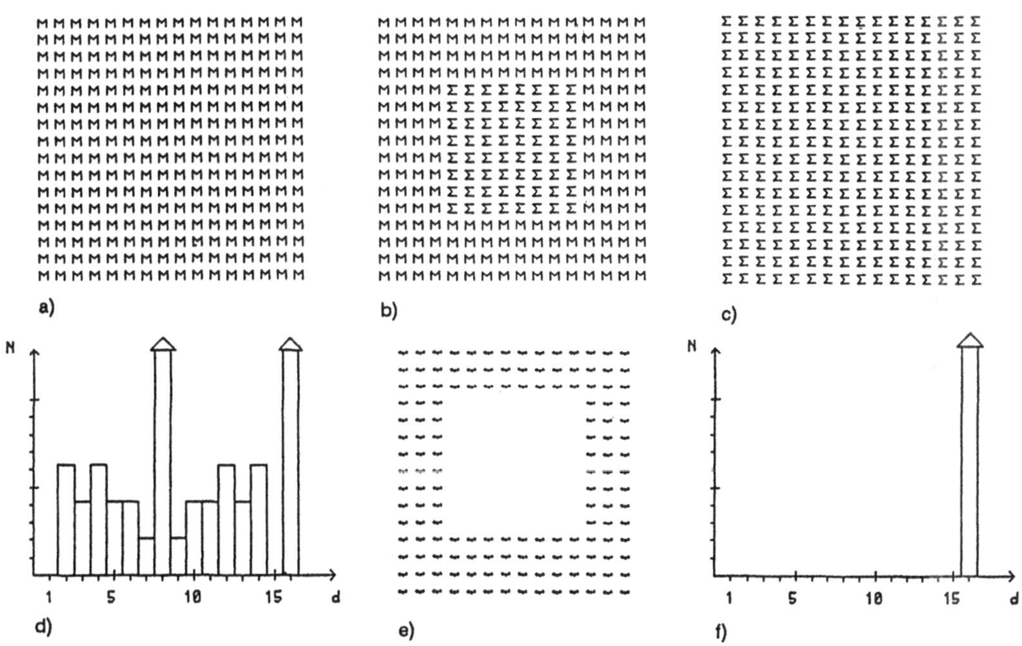

Fig. 3: a) Texture A. b) Texture pair A/B. c) Texture B. d), f) Interval histograms for textures A and B (horizontal scan direction). Intervals between black/white transitions in pixels. (↑ indicates: values are exceeding the display range.) e) Spatial distribution of horizontal intervals of length d = 2 pixels in the texture pair A/B.

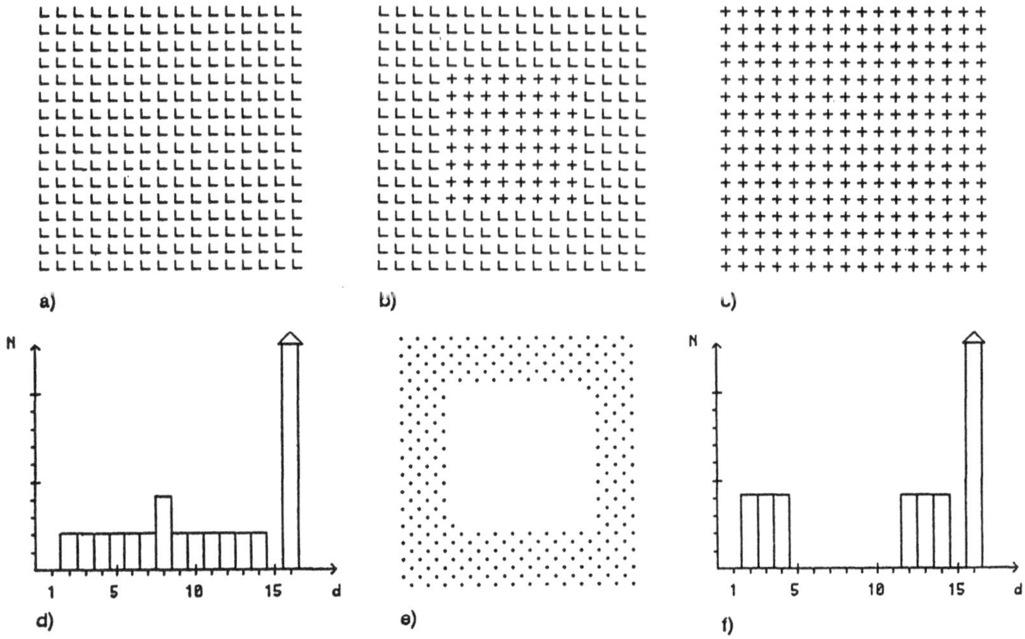

Fig. 4: a) Texture A. b) Texture pair A/B. c) Texture B. d),f) Interval histogram for textures A and B (45° diagonal scan direction). Intervals between black/white transitions in pixels. (↑ indicates: values are exceeding the display range.) e) Spatial distribution of diagonal intervals of length d = 8 pixels in the texture pair A/B.

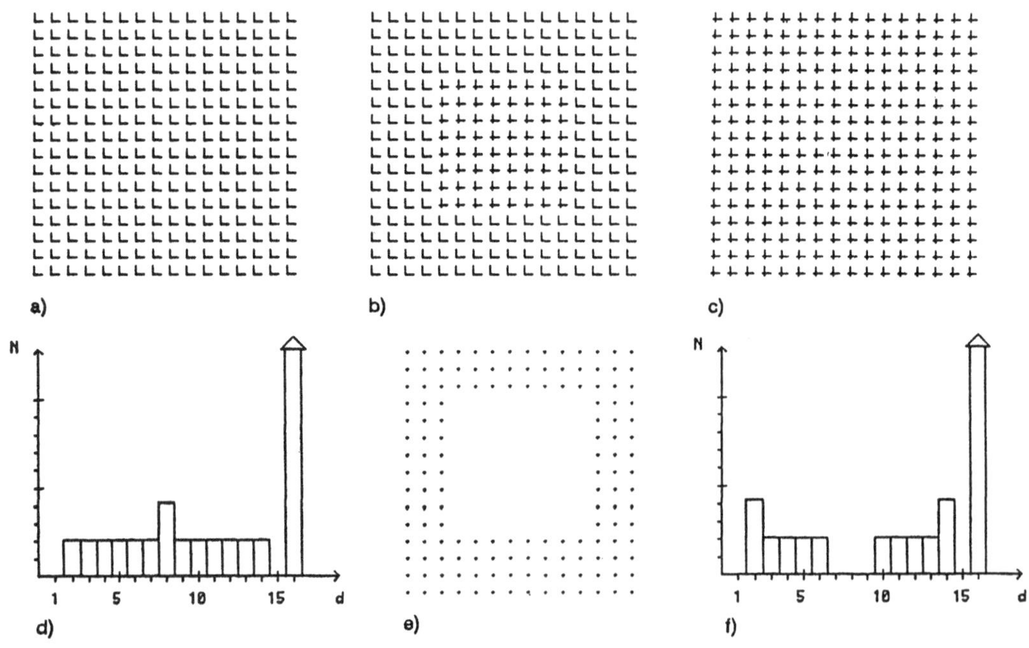

Fig. 5: a) Texture A. b) Texture pair A/B. c) Texture B. d), f) Interval histograms for textures A and B (45° diagonal scan direction). Intervals between black/white transitions in pixels. (↑ indicates: values are exceeding the display range.) e) Spatial distribution of horizontal intervals of length d = 7 pixels in the texture pair A/B.

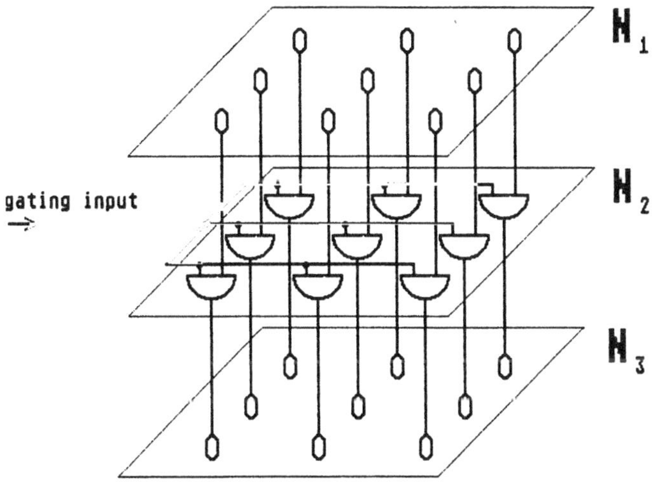

Fig. 6 Texture filtering via a strobed layer (N_2) of AND gates.

interval histograms in Figs. 2d-5d, 2f-5f are for one scan direction only (horizontal in Figs. 2 & 3; 45° in Figs. 4 & 5). Pre-attentively discriminable textures were found to have at least one scan direction that leads to different interval distributions. Additional results on pre-attentive texture discrimination are given in [15, 57]. (It should be noted that the texture pairs Fig. 7 in [15] that were reported to be indiscriminable by our first model (receptive field size 1 pixel) have been shown to be discriminable if the receptive field size is increased. This previous exception is thus also in agreement with pre-attentive texture discrimination by humans.)

The receptive field size in Figs. 2 to 5 has been assumed to be 1 pixel. Considerably more accurate texture analysis would be possible if neurons with specific receptive field structures, similar to those of cortical cells in the primary visual areas, were included in the model. It is surprising, that such excellent agreement with human pre-attentive texture discrimination (and with texture discrimination based on the Texton theory [18-20]) can be achieved with unstructured receptive fields of 1 to 4 pixels, already.

3.3 Stage II: The Linking of "Similarities" by a Hierarchy of Correlators

Starting from the observation that 3 distinct ranges of temporal correlations in the visual cortex (Fig. 8) were found to be related to distinct levels of organization in structure and coding, we are proposing a hierarchical order of mechanisms for the assessment of continuity, for the local and global linking of texture regions, and for object definition.

The structural levels of the second part of our model are organized in several layers of correlation units of increasing complexity. The elements of the lowest layer are single neurons, those of the second layer are local neural assemblies, and the third layer is assumed to consist of a system of interconnected distributed assemblies. Further stages can be added. Each structural level has its specific spatial and temporal interaction ranges within the hierarchic organization: The correlators with high spatial and temporal precision feed their signals to the medium range correlation units and these in turn signal to the next layer with the broadest spatial and temporal interaction ranges. The higher level structures have excitatory feedback loops to the lower levels so that they can support the lower level correlation systems by sensitizing them.

Horizontal "in-layer" connections might be added at each stage: excitatory connections can support the positive feedback from higher layers for "tying together" network elements that are activated by similar stimulus properties. Inhibitory connections might serve to stop signal spread and define borders for region separation.

Additional correlations are introduced by overlapping input ranges (such as overlapping receptive fields or overlapping orientation tuning curves) of the correlator units. These different stimulus-related and internal sources of correlated, i.e. redundant, information depend on the actual visual pattern and on the state of the system; non-redundant information has little chance to "ignite" a substantial part of the correlator network.

The following functional implications might be attributed to the different levels of the correlator network: The narrow-range units are activated by precisely correlated signals generated by local details of a visual pattern. The medium-range units establish signal relations between local groups to combine or separate related figure elements. Broad- range units can exert global region definitions and possibly sensory and cognitive integrations. Such cooperative mechanism of a self-organizing network which can transiently link together neuron groups via correlated activities is a powerful mechanism for region definition and figure/ground separation (see also [6, 14, 15, 54, 57]).

It is of particular interest that reciprocal interconnections between cortical areas and feedback to earlier stages (including the LGN), as they are required in our model, are well documented in the visual system [21-33].

3.4 Experimental Support for a Cooperative Correlator Hierarchy

The proposed correlator model makes certain predictions about the functional modes of the visual system, about neural interconnections, and about the spatio-temporal signal relations. The predicted effects, therefore, should be observable in neurophysiological and psychophysical experiments.

3.4.1 Neurophysiological Evidence

We began to analyze neural cooperative phenomena in cat visual cortex [34-39, 55] using multi-electrode techniques [37, 41-43] together with signal correlation methods [44 - 47]. Neural cooperative behaviour can be studied when correlations due to a common stimulus are separated from those which are due to internal synaptic interactions, by subtracting the appropriate expectation values ("shift predictors" [48, 49]). Multiple recordings and correlation measurements were made from 4 different structural levels of cortex organization: from local clusters of single neurons, single vertical columns, neighboring columns and from different cortex areas.

3.4.2 Stimulus Dependence of Effective Neural Coupling

Visual activation can leave internally generated signal correlations unchanged or (what is more interesting) can change its strength or even its type within milliseconds. This effect is interpreted as a transient change in the effective coupling within the analyzed neural

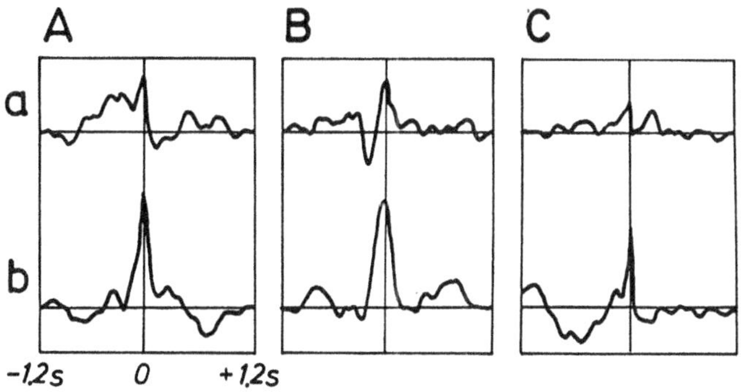

Fig. 7 : Stimulus dependence of effective neural coupling can be shown, when the influence of the common stimulus source is compensated for (subtraction of "shift-predictor"). This has been done for 3 examples of recordings from cat primary visual cortex. Stimulus corrected cross-correlograms of multi-unit activities from the supra- und infra-granular layers of the same vertical cortex column are shown. Stimuli for A: random dot pattern. a: sweep against preferred direction. b: pattern not moved (maintained activity). B: random static dot pattern. a: sweep against, b: in preferred direction. C: optimal bar. a: sweep against, b: in preferred direction. (Experiments in cooperation with R. Bauer and W. Jordan).

network. The activation of different parts of the neural network by different stimuli was specific for the type of stimulation.

In Fig. 7 the stimulus dependence of the effective neural coupling is shown for 3 examples of recordings from cat primary visual cortex. (Stimulus corrected cross-correlograms of multi-unit activities from the supra- und infra-granular layers of the same vertical cortex column).

3.4.3 Distinct Temporal Correlation Ranges

Within and between the different cortical structures distinct correlation ranges could be separated (Fig. 8):
Precise couplings (0-5ms) were found in single unit signals within short recording distances. Correlations of medium width (5-50ms) were apparent at short and medium recording distances a.) between supra- and infra-granular layers of the same cortical column, b.) between neighboring columns and c.) between visuo topically equivalent positions of different areas; here, broad correlations can appear too. Broad correlations (50-500ms) were found within a wide range of electrode distances from nearby to remote recording positions. Due to feedback from higher levels, medium and broad correlation ranges might also be present at small recording distances.

These combinations of spatial and temporal correlation ranges are in good agreement with the postulated levels of cortical assembly organization in our model of hierarchically organized correlators.

4. Psychophysical Support for the Model

It is well known that a region is greatly accentuated if it moves relative to the background. Even regions of identical texture (e.g. random dot patterns) become discriminable if one region is shifted against the other [10, 17, 50, 51]. In addition, the proposed correlator

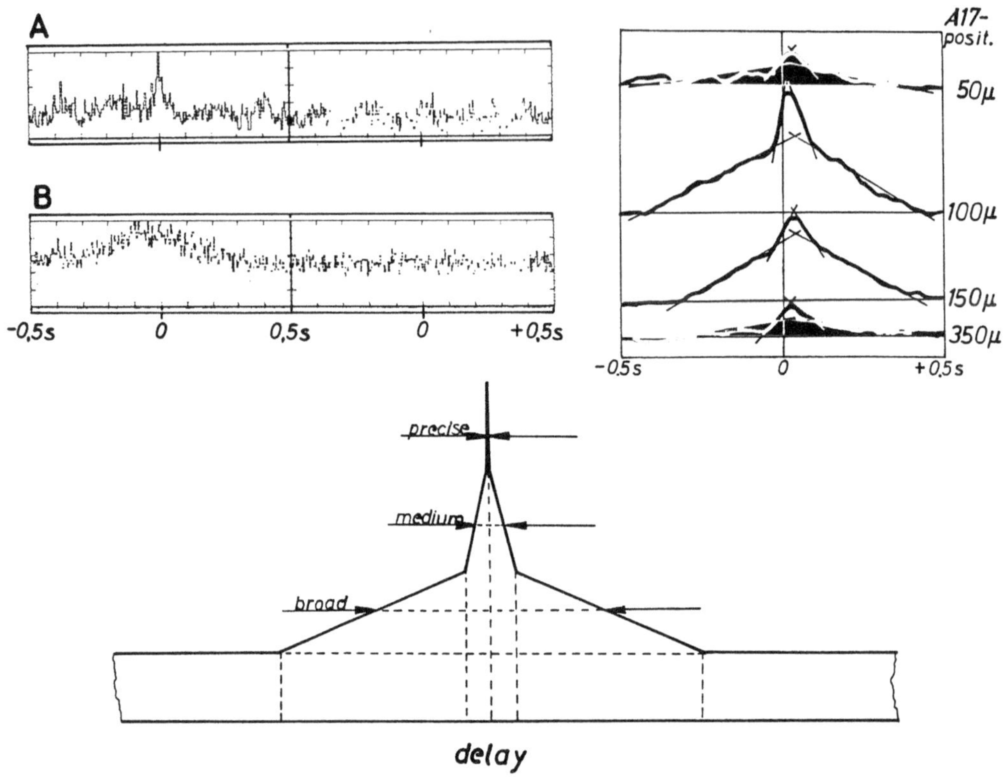

Fig. 8 : Distinct temporal correlation ranges were found under a variety of recording and stimulus conditions. This is summarized in the central scheme; it shows a correlogram with 3 different superimposed ranges of synergistic activation: precise (0-5 ms), medium (5-50 ms) and broad (50-500 ms) correlation width (width at half peak amplitude). Top left: Cross-correlograms from single-unit recordings of cat primary visual cortex; A: nearby electrodes (0.24 mm): precise and medium components are present; B: more distant electrodes (1.1 mm): only broad component is present. Left: Correlograms from data in original order. Right: Isolation of stimulus-induced correlation. (Experiments with R. Bauer and W. Jordan). Top right: Cross-correlograms from multi-unit signals recorded from about iso-topographic representations in 2 visual areas (A17 and A18; A18-position was fixed, 4 different A17-positions). Medium and broad range correlation components are visualized by straight lines. (Recordings with R. Leferink).

model is supported by the following psychophysical results: WILSON and SINGER [52] studied the phenomenon that simultaneous events tend to be perceived as having a common origin. They found that the temporal appearance of a stationary stimulus disk, which was modulated in its intensity, was affected by relatively remote mask stimuli. Their experiments suggested that there exists a perceptual mechanism that associates physically independent events. This "perceptual connection" is effective over 20 ^{0} of visual angle and over target-to-mask onset asynchronities up to 150 ms. Our model of hierarchical correlators with feedback can explain these results: The two remote stimuli form two discrete temporarily connected assemblies at the lower level, which can be "unified", when their outputs a.) converge on higher order elements (which have broad interaction ranges) and b.) when their signals are sufficiently correlated (i.e. similar in their time course).

The ability of the visual system to link temporally dispersed figure elements in dynamic background noise was analyzed by ALTMANN et al. [53]. Their experiments were based on the concept of an MS thesis in our group [56] which in turn was based on ideas of one of us (H.J.R.): The results indicated that the detection of figures hidden in noise is possible even if figure elements are dispersed in time over several hundred milliseconds. It is further suggested that the measured "persistence" must occur at a level of processing where retinotopy is still preserved. Our model of transiently connected correlation units can explain these results: the upper layer with its broad spatial and temporal correlation ranges is able to link broadly correlated events over several hundred milliseconds by establishing corresponding neural interconnections.

5. Summary

Correlated pattern-specific neural activities can be generated and separated by the mechanisms in the first stage of our proposed model. The essential mechanisms of the second stage (the hierarchy of parallel correlation systems with feedback) is that near-simultaneous neural activities can transiently link together parts of the neural correlator network. Furthermore, this network would provide a means to generate from these implicit representations explicit hierarchical codes for region definition and region separation at a higher level of abstraction.

Literature

1. D. Casasent, D. Psaltis: Opt. Commun., Vol. 17, 59 (1976)
2. G. West, H. J. Reitboeck: Elektr. Inform. & Kybernetik, 15, 507 (1979)
3. J. Altmann, H. J. Reitboeck: IEEE, PAMI-6, 46 (1984)
4. H. J. Reitboeck, J. Altmann: Biol. Cybern. 51, 113 (1984)
5. W. Reichardt: In Sensory Communication, ed. by W.A. Rosenblith. (MIT Press 1961)
6. C. von der Malsburg: Internal Report 81-2 (Max-Planck-Institute for Biophysical Chemistry, Goettingen, 1981)
7. P. Johannesma, A. Aertsen, H. van den Boogaard, J. Eggermont, and W. Epping: In Brain Theory, ed. by G. Palm and A. Aertsen (Springer, Berlin, 1986), p. 25
8. W.A. van de Grind, J.J. Koenderink, and A.J. van Doorn: Vision Research, 26, 797 (1986)
9. J.J. Koenderink: In Limits in Perception, ed. by van Doorn, W.A. van de Grind, and J.J. Koenderink (VNU Science Press, Utrecht, 1984) p. 495
10. W. Reichardt: In Pattern Formation by Dynamic Systems and Pattern Recognition, ed. by H. Haken (Springer, Berlin, Heidelberg, 1979) p. 100
11. W. Reichardt, T. Poggio: Biol. Cybern. 35 (1979)
12. W. Reichardt, T. Poggio, K. Hausen: Biol. Cybern. 46, 1 (1983)
13. M. Egelhaaf, W. Reichardt: Biol. Cybern. 56, 69 (1987)
14. H.J. Reitboeck: In Synergetics of the Brain, ed. by E. Basar, H. Flohr, H. Haken, and A.J. Mandell (Springer, Berlin, Heidelberg 1983) p. 174
15. H.J. Reitboeck, M. Pabst, R. Eckhorn: In Computer Simulation in Brain Science, ed. by R.M.J. Cotterill (Phys. Scripta, Cambridge University Press, Cambridge 1987)
16. R. Eckhorn, B. Poepel: Vision Res. 21, 1693 (1981)
17. B. Julesz: Foundations of Cyclopean Perception, Univ. of Chicago Press (1971)
18. B. Julesz: Nature 290, 245 (1981)
19. B. Julesz: Biol. Cybern. 41, 131 (1981)
20. B. Julesz: Biol. Cybern. 54, 245 (1986)
21. B.G. Cragg, a. Ainsworth: Vision Res. 9, 733 (1969)
22. S.M. Zeki: Brain Res. 14, 271 (1969)
23. S.M. Zeki: Proc. R. Soc. Lond. B 207, 499 (1980)

24. J. Tigges, W.B. Spatz, M. Tigges: J. Comp. Neurol. 148, 481 (1973)
25. J. Tigges, W.B. Spatz, M. Tigges: J. Comp. Neurol. 158, 216 (1974)
26. M. Wong-Riley: Brain Res. 147, 159 (1978)
27. K.J. Rockland, D.N. Pandya: Brain Res. 179, 3 (1979)
28. L.G. Ungerleider, M. Mishkin: J. Comp. Neurol. 188, 347 (1979)
29. J.S. Lund, A.E. Hendrickson, M.P. Ogren, E.A. Tobin: J. Comp. Neurol. 202, 19 (1981)
30. H.R. Maunsell, D.C. van Essen: J. Neurosci. 3: 2563 (1983)
31. R.E. Weller, J.H. Kaas: J. Comp. Neurol. 220, 253 (1983)
32. L.G. Ungerleider, R. Desimone, M. Mishin: Soc. Neurosci., Abstr. 8, 680 (1982)
33. L.G. Ungerleider, R. Gattass, A.P.B. Sousa, M. Mishkin: Soc. Neurosci., Abstr. 9, 152 (1983)
34. R. Eckhorn, C. Habbel, F. Krause, H. Lohmann, H.J. Reitboeck, J. Schneider: Neurosci. Lett. (Suppl.) 18, 164 (1984)
35. R. Eckhorn, J. Schneider, R. Keidel: J. Neurosc. Meth. 18, 371 (1986)
36. R. Eckhorn, H.J. Reitboeck: In Dynamics of Sensory and Cognitive Processing of the Brain, ed. by E. Basar (Springer, Berlin, Heidelberg, New York 1987)
37. H.J. Reitboeck, W. Adamczak, R. Eckhorn, P. Muth, R. Thielmann, U. Thomas: Neurosc. Lett., Suppl. 7, 148 (1981)
38. J. Leferink, R. Eckhorn, J.I. Nelson, H.-J. Wagner: Submitted to IBRO, 2nd World Congress of Neuroscience, Budapest (1987)
39. R. Eckhorn, R. Bauer, W. Jordan, W. Epping, M. Arndt: Submitted to IBRO, 2nd World congress of Neuroscience (1987)
40. H.J. Reitboeck, K. Konkol: Biophys. Struct. Mechan., 7, No. 4, 342 (1981)
41. H.J. Reitboeck: IEEE Trans. SMC-13, No.5, 676 (1983)
42. H.J. Reitboeck: J. Neurosc. Meth. 8, 249 (1983)
43. H.J. Reitboeck, G. Werner: Experimentia 19, 339 (1983)
44. J. Schneider, R. Eckhorn, H.J. Reitboeck: Biol. Cybern. 46, 129 (1983)
45. J. Schneider, R. Eckhorn, R. Keidel: Europ. J. Physiol. 402, R56 (1984)
46. J. Schneider, R. Eckhorn: Neurosci. Lett. 18, 164 (1984)
47. R. Eckhorn, J. Schneider, R. Keidel: J. Neurosc. Meth. 18, 371 (1986)
48. G.L. Gerstein, D.H. Perkel: Biophys. J. 12, 453 (1972)
49. J. Schneider, R. Eckhorn, H.J. Reitboeck: Biol. Cybern. 46, 129 (1983)
50. O.J. Braddick: Vision Res. 14, 519 (1974)
51. C.L. Baker, jr., O.J. Braddick: Vision Res. 22, 851 (1982)
52. G.T.L. Wilson, W. Singer: Perception & Psychophysics, 30, 107 (1981)
53. L. Altmann, R. Eckhorn, W. Singer: Vision Res. 26, 1949 (1986)
54. C. v. d. Malsburg: In Synergetics of the Brain, ed. by E. Basar, H. Flohr, H. Haken, and A.J. Mandell (Springer, Berlin, Heidelberg 1983)
55. R. Eckhorn, H. Lohmann, H.J. Reitboeck: ENA Satellite Vision Congress, Lyon, France (1986)
56. L. Altmann, M.S. thesis, Marburg/Hamburg (1982)
57. H.J. Reitboeck: In Organizing Principles of Sensory Processing: The Brain's View of the World, ed. by J.S. Lund (Oxford University Press, New York 1987)

Acknowledgements

This work was supported by grants from Deutsche Forschungsgemeinschaft (Re 547/2-1, Ba 636/4-2) and from the National Institutes of Health, U.S.A. (1PO1 NS 20868).

Self-organization of the Visual Information Channel and Solitons

R.S. Ingarden

Institute of Physics, N. Copernicus University,
PL-87-100 Toruń, Poland

1. INTRODUCTION

Human vision is a very complicated dynamical process in which not only the eye, but also the brain, as well as other sense organs take active part. Even such a seemingly constant property as the transparency of the cornea is a result of an intensive metabolic activity causing a circulation of water in this tissue. After death the circulation stops and the transparency disappears. When we stare unmovingly at a constant scene, it seems that our eyes do not move, but actually they do inappreciably in fractions of a second (the saccadic movement /1/). By these movements the eyes scan constantly the field of view which otherwise would be of the order of 1° only (this is the angular diameter of the macula, the fovea centralis, where the cones are concentrated, the only light receptors active by day illumination). Our actual field of view is of the order of 80° or more. The coordination and analysis of the impressions of the light receptors is done by the complicated network of many neuron layers (3 layers in the retina, 6 in the lateral geniculate body, and 6 in the primary visual cortex + connections with many other parts of the brain, especially, Wernicke's area, inferotemporal cortex, superior colliculus and frontal eye fields, the latter for the saccadic movements, cf. /2/-/6/). The primary visual cortex is mainly responsible for pattern recognition and the superior colliculus for localisation in space in coordination with hearing and the equilibrium sense /2/. The scale of illumination is automatically adjusted to some mean intensity typical for our seeing, as if the objects were illuminated by ourselves. (This was probably the reason why the ancients imagined the light rays as coming from the eye and not to the eye.) All this occurs because our biological aim is to see objects and not only the light. We are born "realists", sometimes even "surrealists" or "abstractionists", but never "impressionists" in our natural vision attitude. Therefore philosophically, in the theory of vision the saying of the young Husserl "zurück zur Sache" is very proper, while his idea of "phenomena" ("phenomenology") is rather confusing and contradictory. In physics and physiology we do not seek "phenomenas" but "things", "objects", "states", "reality".

The purpose of the present preliminary note is to try to formulate the mathematical principle of vision as a variational principle of special type. The completion of this idea requires still a lot of special experimental information to obtain a concrete mathematical model of vision. Nevertheless, the author hopes that the general idea can help to find in future such a concrete model.

2. INTENSITY OF LIGHT AND INTENSITY OF LIGHT PERCEPTION

Up until recent times people imagined the vision process as something similar to the photographic process, cf. /7/, p. 357. How different, however, the two processes are can be seen from 2 photographs presented in Fig. 1. They were taken in Japan last year in my room (in the guest house of the Science University of Tokyo) in the same illumination conditions by means of an automatic Japanese camera. The first photo was taken, however, in the direction of the west window, where the sun was just shining (it was afternoon), while the other in direction of the east

a

b

Fig.1. The same room in two directions. a) In the direction of the sun, b) in the opposite direction

window not illuminated by the sun. Subjectively, the room was perceived as equally well illuminated and bright (the electric lights were also burning, as is seen in the first photo). But the camera, which adjusted its illumination by an automatic aperture change, has seen the conditions differently to the eye, and differently in the two opposite directions.

The physical reason for such a big difference in photographic and visual recording lies in the different mechanisms of representation of the light intensity. On the photographic negative a higher light intensity means a greater density of silver grains. In the nerve fibre, however, we may have only quicker or slower time series of the same spikes of excitation, "yes" or "no". The frequency of spikes varies from 0 to ca. 1000 per second. But second is here an arbitrary unit, we have to consider rather a "perception time" of the order of 150 ms (milliseconds). It corresponds to the time of persistence of the image (e.g. in motion pictures) and has the sense of the visual relaxation time. The same is the period of the saccadic movement, in a shorter time we would not see a new portion of the field of view. Thus in our visual perception we have not more than ca. 150 levels of intensity and pictures cannot change in time shorter than ca. 0.15 sec. We have also a special mechanism for motion recognition in the primary visual cortex, cf. /2/ - /6/, which enables us to perceive much quicker motions and their direction, but this does not mean that all the picture changes so quickly. We only perceive that something changes in a given direction, but do not see the details yet.

But for adjusting the level of intensity and the contrast of the picture, another phenomenon is crucial. Namely, the neurons can produce and adjust some "background excitation" as a "zero level" of our perception, cf. /2/. With respect to this level (frequency of spikes) a lower frequency is felt as an "inhibition" and a higher frequency as an "excitation" (one considers also dendrites with positive and negative gains). We give here an example from /2/, p. 159 of the German edition, where are 5 input dendrites with gains -1/8, 1/2, 1, 1/8, and -1/2, respectively, the background is assumed as 100 and inputs as 80, 78, 75, and 70, respectively. We obtain then the output in the axon of the neuron as equal to (the units are arbitrary)

$$100 - \frac{1}{8}\,80 + \frac{1}{2}\,78 + 75 + \frac{1}{8}\,72 - \frac{1}{2}\,70 = 178 \quad . \tag{1}$$

We see that the output essentially depends on the background, and the latter should be always such that the output is nonnegative. If we consider the gain coefficients of the dendrites as given by their anatomy, the only physiologically free parameter is the background excitation of the neuron. If so, it seems that this parameter adjusted sufficiently for each neuron can explain the global "normalization" of the intensity of excitation, $v(x) \geqslant 0$, $x \in X$, over the visual field of view $X = \{(x_1,x_2)\}$, approximatively,

$$\int_X v(x)d^2x = 1 \quad . \tag{2}$$

It is interesting that the range of light intensities perceived by the eye and the range of sound intensities perceived by the ear are in relative scale almost the same, ca. 10^{16} or 160 dB (decibels). It seems that this fact confirms our hypothesis of the neural origin of the optical adaptation, probably the same in principle for the acoustical adaptation.

3. RECEPTIVE FIELDS

Stephen W. Kaffler discovered in 1953 that the ganglion cells of the retina (and analogously the cells of the layer IV of the primary visual cortex) have concentric receptive fields, i.e., the sets of receptors (cones) which may cause their excitation. The fields are of two types: ON-centre/OFF surround, and OFF-centre/ON surround /2/ /5/ /6/. That means that when the receptors are excited by light, the central or peripheral parts of the receptive field act excitingly, while the other part inhibitory. As we have mentioned above, the inhibitory action depends essentially on the background excitation of the secondary cell. It can be easily seen, cf. Fig. 2 taken from /2/, that such a structure of the neural network can cause a contrast increase in the final image. Thus the same mechanism as that of the overall daylight adaptation described in Section 2 also causes (by special anatomical structure) an adjustable change in the contrast of the visual picture, very important for the perception of objects and the pattern recognition. With respect to the real optical image on the retina, we see therefore much stronger contours of objects, like in the black lines drawn by the XIX-th century painters. (These black lines were strongly criticised by the impressionist painters as being against nature and true vision, actually they were against true nature, but not against true vision).

4. THE FIRST MATHEMATICAL FORMULATION

In my younger years (1954-1960) I wrote a series of papers on optimal optical systems /8/-/11/. In /8/ /9/ the classical methods of geometrical and diffraction optics have been used, also in a generalized and combined form, while in /10/ /11/ the ideas of information theory have been applied and a generalization of variational calculus has been proposed. With usual instruments we have the difficulty that the optimization can be performed, strictly speaking, only for a given object and illumination. That is the reason why the modern optical manufacturers, Japanese for instance, produce so many types of lenses for the same camera, the number of such objectives being of order of 50 or more. But a living optical system, the eye, is able to optimize its performance for any given type of object and illumination. For this purpose the eye has its mechanism of adaptation (to the illumination) and accommodation (to the objects, their position and shape). Independently of the head and eye movements there are also movements of the iris, of the lens and

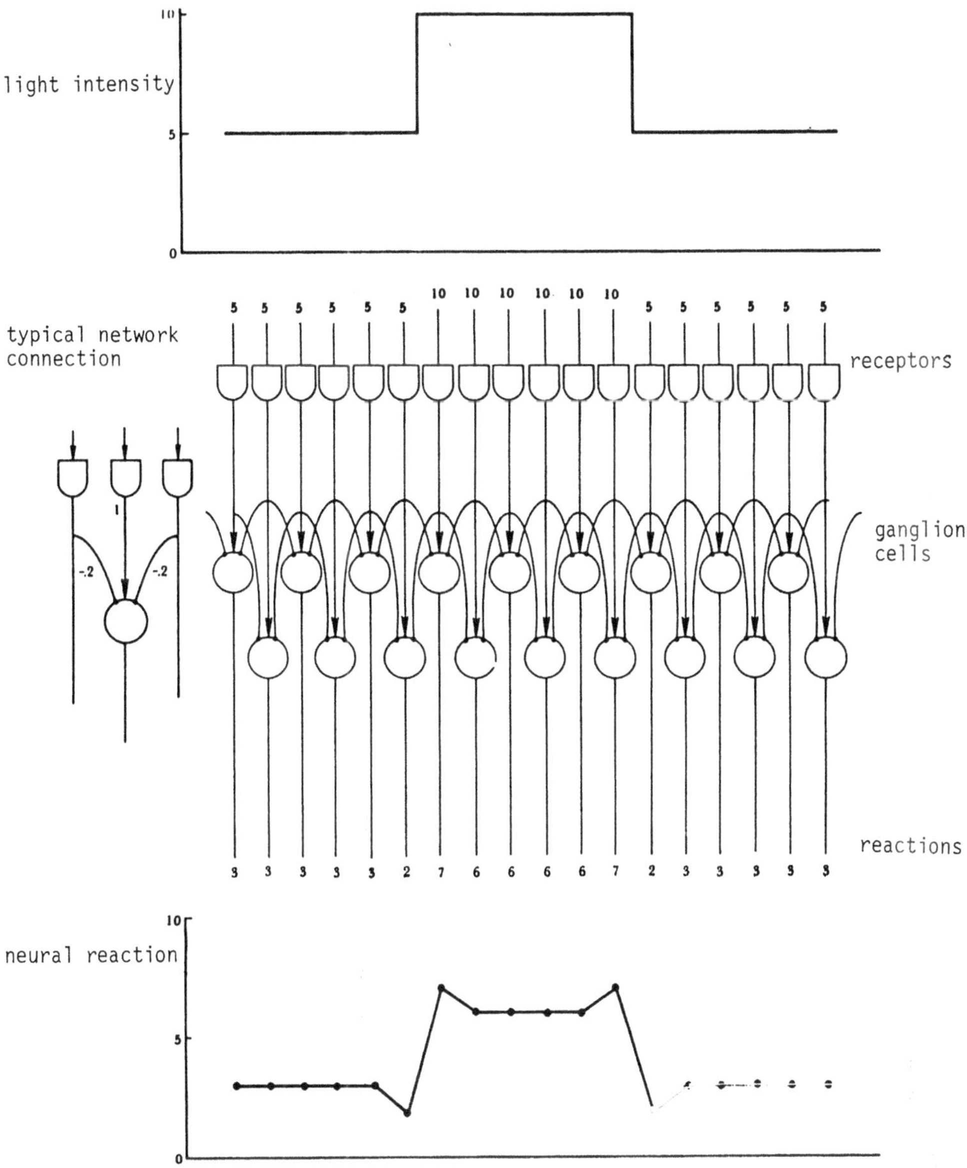

Fig. 2. Contrast increasing as a result of ON-centre/OFF-surround receptive fields in the neural network

changes of the background excitations of neurons in the neuron network, mentioned above in Sections 2 and 3. Such processes have the character of feedback control and self-organization. They belong to the field of synergetics, if we understand it in a wide sense, i.e., of common <u>action</u> of many elements, not necessarily of common energy, rather of common information or entropy (synentropy, but such a term may be unnecessary since Professor Haken's intention in the definition of synergetics seems to be rather very wide, not restricted to energy only).

Let us denote the coordinates in the object plane before the eye be $y = (y_1, y_2) \in Y$, the light intensity in this plane (by non-coherent monochromatic illumination) by $u(y)$, and the corresponding quantities after processing by the system eye + brain (i.e., excitations seen by us in the vision process, disregarding however the color vision, the stereoscopic vision, perception of motion and similar higher order perceptions) by $x = (x_1, x_2) \in X = Y$ and $v(x)$. Roughly speaking we may say that $v(x)$ describes the frequencies of excitation spikes on the neurons of the layer IV of the primary visual cortex, integrated in the process of saccadic motions to one picture. The global result of the visual transformation (not only optical, but combined optical + neural) can be represented by a non-linear Gibbs formula of the form of a Hammerstein operator, cf. /12/ p. 361,

$$v(x) = \int_Y K(x,y) f(y,u(y))\, d^2y = \underline{Kf}\, u \quad , \tag{3}$$

where $\underline{K}$ is a symmetric, positive definite Fredholm operator, $K(x,y) = K(y,x)$,

$$K(x,y) \geq 0 \text{ for } x \in X, y \in Y, \int_X \int_Y K^2(x,y) d^2x d^2y < \infty \text{ (e.g., = 1)} \quad , \tag{4}$$

and $\underline{f}$, called a superposition operator, is a non-linear positive, smooth function of $y \in Y$ and $u \in [0,\infty]$ describing the feedback of the system, cf. /13/, and satisfying the inequality

$$f(y,u) \leq a + b\, u, \ a,b \text{ - constants} \quad . \tag{5}$$

Under these assumptions the eigenvalue problem

$$\lambda u = \underline{Kf}\, u \tag{6}$$

has a continuum of solutions, cf. Theorem 2.23 of /12/, p. 402.

As a simple example let us consider

$$K(x,y) = A(|x - y|) \geq 0, \int_X A(|x|) d^2x < \infty \ , \ f(y,u)$$

$$= \frac{u}{1+a(y)u} \ , \ a(y) \geq 0 \quad . \tag{7}$$

For our biophysical applications we are not interested, however, in the "trivial" solutions (6), but in solutions which increase the sharpness and contrast of the image with respect to the original object, i.e., "surealistic" ones, with higher information. For such a purpose we first try the following double variational principle:

$$S(v,u) = \int_X v(x) \ln\left(\frac{v(x)}{u(x)}\right) d^2x = \min, \quad (8)$$

$$S(v,1) = \int_X v(x) \ln(v(x))\, d^2x = \max. \quad (9)$$

(8)–(9): mean value (cf. below)

under assumptions (X = Y)

$$\lambda\, v(x) = \underline{Kf}\, u(y), \quad \int_Y u(y)\, d^2y = 1, \quad \int_X v(x)\, d^2x = 1, 0 < \lambda < \infty \quad . \qquad (10)$$

In this principle function u(y) is considered as given, while operator Kf as unknown, to be fixed. We factorized this operator into operator K representing the optical part of your system (the eye) and operator f representing the neural part (the brain). So optimizing of K corresponds to the accommodation of the lens and optimizing of f to the adaptation of the neural cells. In the "mini-max" principle (8)-(10) the minimum of S(v,u) describes the similarity of functions v and u, of the image and of the object, and the maximum of S(v,1) the dissimilarity of v and homogeneous plane without any "picture". In both cases the criteron is the relative amount of information given by the Kullback integral, called a Kullback (or Kullback-Leibler) information, cf. /14/-/16/.

Minimum (8) and maximum (9) are taken with respect to the same class B of representation operators Kf, Kf ∈ B, defining the model. E.g., in example (7) this class is defined by two control elements: $A(|x|) \in L^1_+(0,\infty)$, $a(y) \in L^1_+(X)$. We have to fix the exact meaning of this principle since without such explanation condition (8) may contradict condition (9). In general, operators Kf minimizing (8) or maximizing (9) may not exist, and if they exist, they are different, as a rule. So we have to make some compromise between the two conditions, namely, the mean value of the resulting operators, if they exist. We understand this mean value in the sense of the functional spaces of the corresponding control elements, e.g., in the simplest case of control parameters, the numerical mean value of these parameters. We explain this procedure in the following simple example.

5. AN EXAMPLE

We consider example (7), but for the sake of simplicity only its very special case. Since we are mostly interested in the neural mechanism of vision, we assume the optical part as ideally representing,

$$A(|x|) = \delta\,(x), \; K(x,y) = \delta(x - y) = \delta(x_1-y_1)\,\delta\,(x_2-y_2) \quad , \qquad (11)$$

and control function a(y) as a control parameter a, $0 < a < \alpha \ll 1$.
Thus

$$v(x) = \frac{u(x)}{1+a\,u(x)}, \quad \int_X v(x)\, d^2x \approx 1. \qquad (12)$$

We denote as

$$A = \int_X u^2(x)\, d^2x, \quad B = - \int_X u(x) \ \ln(u(x))d^2x \quad , \tag{13}$$

the square and logarithmic informations of u (x), respectively, assuming that they exist. We obtain then from (8) and (9)

$$\min_{0 \leq a \leq \alpha} (- aA) \to a = \alpha \ , \quad \max_{0 \leq a \leq \alpha} (- B - fa) \to \alpha = 0 \ , \quad a_0 = \frac{\alpha}{2} \ , \tag{14}$$

cf. Fig. 3.

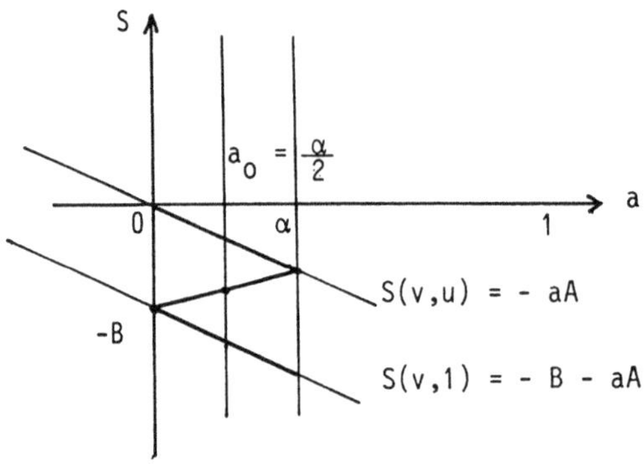

Fig. 3. An example of the "mini-max" optimization principle

6. A GEOMETRICAL INTERPRETATION AND A NEW FORMULATION

In the introduction we pointed out that vision is a dynamical process, i.e., a process taking part in some time, the visional relaxation time, actually of a few saccadic movements of the eye, approximately half a second. In our above example this may roughly correspond, e.g., to the zigzag curve presented in Fig. 3. But there we have no time, actually. So it will be better to introduce time in some way. This is possible by means of a geometrical interpretation of thermodynamcis by means of Finsler geometry, cf. /17/ /18/. In this interpretation in the differential manifold of thermodynamical states (in our case functions v(x), u(x) correspond to these states) a non-isotropic metric is introduced by means of relative entropy (8). To get a homogeneous Lagrange function of the geometry, instead of a parameter along a geodesical line, a new dimension is introduced, called time. This is done in the lowest nontrivial Taylor approximation of the relative entropy, as a local metric. In such an interpretation we may imagine a bundle of geodesics coming out from the initial state, u(x). They all realize minimization of relative entropy (8) and lead to a final state v(x) after a relaxation time of the vision process. To choose between all the possible geodesics we need another condition, and this is just the maximum condition (9) for v(x). In such a way we see the necessity of both conditions (8) and (9), as well as the time of the process which is included in the interpretation.

7. SOLITONS

In order to go over from our proposition to an actual model of vision we need more realistic assumptions of the class B than our very simplified model. If we use Finsler geometry, the theory is nonlinear and highly complicated. It seems that perhaps the Finsler model may be assumed to be such that the solutions of the field or geodesic equations be completely integrable, i.e., in the form of soliton-like solutions. Such solutions have been already obtained in the same special Riemannian spaces which are in principle very similar to the Finsler ones. In such a case we would have many-soliton solutions of very simple properties, as is actually possible in the case of vision. In biology it is well known that solitons play an essential part in muscle and nerve processes, so it is not excluded that they may be also useful in mathematical modelling of the vision process. So far it is, however, only a hope, as well as a proposition for future research.

REFERENCES

1. A.T. Bahill, L. Stark: Sakkadische Augenbewegungen. In Wahrnehmung und visuelles System (Spektrum der Wissenschaft, Heidelberg 1986) p. 68-77
2. P.H. Lindsey, D.A. Norman: Human Information Processing, 2nd ed. (Academic Press, New York), German transl. (Springer, Berlin 1981)
3. M. Imbert: The Coding of the Images in the Retina, Europhysics News 18, 32-36 (1987)
4. O.D. Crentzfeldt: Informationsübertragung und -verarbeitung im Nervensystem. In Biophysik, ed. by. W. Hoppe et al. (Springer, Berlin 1977) p. 629-652
5. C.R. Noback, R.S. Demarest: The Human Nervous System (McGraw-Hill Intern., Singapore 1981), Chap. 12, The Optic System, p. 388-424
6. D.H. Hubal, T.N. Wiesel: Die Verarbeitung visueller Information. In Wahrnehmung und visuelles System (Spektrum d. Wiss., Heidelberg 1986) p.36-47
7. E.R. Kendel, J.H. Schwartz (eds.): Principles of Neural Science, 2nd ed. (Elsevier, New York 1985)
8. R.S. Ingarden, H. Ochman: Optimal Optical Systems, Bull. Acad. Polon. Sci. Cl. III 2, 231-276 (1954)
9. R.S. Ingarden: A Generalization of the Young-Rubinowicz Principle in the Theory of Diffraction. Acta Phys. Polon. 14, 77-91 (1955)
10. R.S. Ingarden: Composite Variational Problems. Bull. Acad. Polon. Sci. Ser. math. 7, 687-689 (1959)
11. R.S. Ingarden: The Problem of Optimum Optical Systems. Optica Acta 7, 179-183 (1966)
12. Zabreyko et al.: Integral Equations - a Reference Text (Noordhoff Intern. Publ., Leyden 1975)
13. J.C. Willems: The Analysis of Feedback Systems (The MIT Press, Cambridge, Mass. 1971)

14. J. Kullback, R.A. Leibler: Information and Sufficiency, Annals Math. Statist. 22, 79-86 (1951)

15. S. Kullback: Information Theory and Statistics (Wiley, New York 1959)

16. S. Gujasu: Information Theory with Applications (McGraw-Hill Intern. Publ., New York 1977)

17. R.S. Ingarden: Geometry of Thermodynamics. In Proc. XV Intern. Conf. Diff. Geom. Methods in Theor. Phys. (H.D. Doebner ed.). Clausthal 1986 (World Scientific, Singapore 1987)

18. R.S. Ingarden: On Physical Interpretations of Finsler and Kawaguchi Geometries and the Barthel Nonlinear Connection. Tensor, N.S., to be published

Human Brain EEG Fields: Micro-states and Their Functional Significance

D. Lehmann, D. Brandeis, H. Ozaki, and I. Pal

Neurology Department, University Hospital, CH-8091 Zürich, Switzerland

1. Functional States of the Brain

Human behavior in response to received information depends on the person's general condition, and in particular on the brain functional state which existed at the moment when the information arrived; likewise, "spontaneous" behavior (volontary or automatic acts) depends on the state during which the act is executed. Not only overt behavior, but also subjective experiences such as perceptions, thoughts, ideas and emotions vary with the momentarily present functional state: The existing state permits and constrains the different brain functions. Well-known examples are the different speeds and reliabilities of motor reactions to tone stimuli in wakefulness and sleep, and the different types of thoughts and spontaneous ideas associated with these states: abstract and reality-referred thoughts vs associative, reality-remote concatenations of thoughts and dream-like mentations (see /1/); related differences are observed during wakefulness between adults and children. Minor but distinct changes of short-lasting functional states have also been studied extensively. The classical "fluctuations of attention" /2/ actually refer to spontaneous variations of perceptual threshold which occur in the range of seconds.

2. State-Dependent Higher Brain Functions

Perceptual threshold is only one of several key parameters which are state dependent; studies along a dimension of vigilance have shown that physiological parameters are correlated with state-dependent, general thought strategies as mentioned above /1,3/, and with state-dependent storage of information and in particular state-dependent access to memory material stored earlier /4/. The state dependent memory retrieval function is of special interest because of its asymmetry: Material stored in higher-order (more developed, more complexly organized) states can be retrieved during lower states, but material stored during lower states becomes inaccessible for higher states. Emotions, which might most simply be viewed as "flags" on information, are state dependent as suggested by studies of sleep onset revery and dreams when emotions tend to be inadequate, possibly reflecting decreased access to body information feedback (heart palpitations, muscle tone, sweat) which is physiologically lower in sleep; emotional changes which have been observed in paralysed (deafferented) patients might be partially caused by the functionally similar changes in physiology.

3. Questions about Characteristics of States

Considering states like wakefulness and sleep, consciousness and unconsciousness, state distinction is obvious, and it appears that if there are intermediate conditions, they are fragile, fleeting and unstable. In conclusion, there seem to be basic, gross, discontinuous states of brain function which are clearly distinguishable. Is there evidence for the idea of a 'non-digital' state continuum on lower levels? We and others have shown close correlations between different types of spontaneous or induced human cognition and various EEG parameters obtained by spectral analysis methods of short (say, 16sec) epochs of EEG immediately preceding the subject's report /1,3,5,6/.

The states of "fluctuations of attention" which were discussed above - they occur as clearly distinguishable states - show much shorter durations. Can this be extended further to a micro level in time and function? Are there at last very brief, basic state units which are still encompassing the large possible variety of brain functions, i.e. which represent global states? To use a catch phrase, are there atoms of thoughts or emotions or consciousness? In this context the principle characteristic of periodicity which is inherent in all life phenomena comes to mind: May global states be installed for extended periods by periodic re-installation?

4. Changes of Brain States

Within the constraints given by an existing functional state, by maturational age /7/, by metabolic state, and by drug and disease influences, the functional state is continuously re-adjusted according to the demands which arriving informations or newly constructed mental conclusions make on the system /1,8/: However, these demands are the result of interpretation of the information by the brain, and the interpretation procedure uses assessment mechanisms (strategies) and criteria (context material which includes meaning and goals) which vary with the momentarily present functional state.

A drastic change of functional state may successfully be implemented within a very short time period, if motivation (past experiences="expert knowledge") is available to judge the state changes as necessary because the information was recognized as novel or important (PAWLOW's "orienting reaction" /8/). For example, the arousal of a sleeping person engaged in dream-type, reality-remote mentation to wakeful, reality-oriented behavior takes little time in case of a convincing fire alarm, but might not work on a routine day.

5. Brain electric activity as manifestation of State

The momentary functional state of the brain is manifest in many different assessment techniques (anatomical, chemical, behavioral, introspective). The present discussion concentrates on measurements of the spatial and temporal configuration of the electric component (Electro-Encephalo-Gram=EEG, Event Related Potentials=ERP) of the electromagnetic brain field as state descriptor. (The EEG signals of interest roughly range from 1 to 200 microV in a band of 0.1 to 40Hz and 5 to 50cm/wave, ERP signals from 0.1 to 30microV in a band of 0.1 to 2000Hz and 5 to 50waves/cm.) The brain electric field (EEG) is recorded via electrodes attached to the scalp by conductive paste. These measurements permit non-invasive, continuous data collection with a very high sensitivity (correlation) to minor changes of brain state (for example, percept of a visual figure reflected in ERP /9/; effect of a single dose of a short-acting sleeping pill detectable two days later in EEG /10/; fluctuations of visual threshold reflected by EEG /11,12/). EEG and ERP also offer a very high time resolution, down to fractions of milliseconds (brain stem evoked potentials). Because of the unavailabiltiy of a unique solution for the "inverse problem" if more than one generator produces the field, the estimation of generator locations in the brain necessitates accepting considerable constraints. Of particular interest are the systematic variations of EEG and ERP measures which are rulefully associated with changes from excitation to relaxed wakefulness to sleep along a dimension of vigilance /13/, where in general increasing amounts of slow frequency electric events occur with decreasing behavioral vigilance.

In spite of the availability of a large variety of data in brain electric field measurements, the data can be related to global, unitary performance characteristics of the brain's functional state such as vigilance or consciousness or processing strategy /1,5,13/. These are global characteristics into which sub-operations are embedded which however are non-existent on the next lower phenomenological level. Nevertheless, the global, functional state might be conceived of as the sum of very many local states which manifest the activities of the various processors and analyzers. In the quest for basic building blocks of brain function, the classification of the momentary electric field as belonging to one of a limited number of classes of states is proposed. If the recorded configuration or landscape of the momentary brain field differs between two moments in time, different neural populations must have been active; it is very reasonable to assume that the activity of different ensembles of neural elements implies different steps or modes or contents of information processing.

6. Assessment of spontaneous brain electric fields (EEG)

Here again the question arises whether there are recognizable, stable epochs of such brain electric landscapes and hence, possibly separable state units of brain information processing, or whether the states change in a continuous manner into each other. Further, traditional time series-oriented EEG work had stressed the apparent spatial propagation or "travelling" of brain electric phenomena, invoking suggestions of continuous processes, but typically has relied on pre-selected points in space as assumedly inactive reference locations. Although recently, adaptive segmentation of the selected EEG voltage-time series aimed at identifying stationary sub-epochs, the spatial information inherent in EEG and ERP data has not been considered adequately /see 14/. Starting with 48-channel recordings in 1969, our lab has developed space-oriented analysis approaches for brain electric field data and used them in EEG and ERP studies on perception and cognition, viewing the brain electric activity data as series of non-ambiguous field maps /14,15,16,17/.

The momentary maps of a series can differ in spatial configuration ("landscape") and in global electric strength or prominence of spatial profile ("hilliness"). The examination of a map series of spontaneous EEG recorded during resting conditions with closed eyes from normal subjects shows frequent, common features: Map hilliness reaches maximal values about 20 times per second.

> Map hilliness /18/ or the related Global Field Power /14,18,19,20/ can be computed as the standard deviation of all momentarily measured potential values. This one-number measurement assesses the degree of hilliness or flatness of the landscape for any number of electrodes. In the resulting unique curve of global field power over time, the maximal points determine maps with optimal signal-to-noise ratios for major features of the landscape.

Selecting the maps at the times of maximal global field power, one notes that successive, selected maps often show similar landscapes of inverted polarity, i.e. that there are about 20 reversals of polarity per second /18,21/. This phenomenon constitutes the classical 9-10Hz "alpha EEG" activity of the traditional voltage-time series.

> The dominant features of the pattern of the momentary landscape can be directly extracted as reference-independent descriptors /9,14,18,19/: The locations of the map's maximum and minimum potential (thus invoking a dipole model), and the point of gravity (the centroid). (Two-dimensional Fourier transforms which have not yet been systematically applied to EEG/ERP map series will likewise require methods of feature or pattern recognition for reduction.)

It can be shown that over time, the mean of the maps (polarity-rectified) at times of maximal global field power is very similar to the mean of all maps (polarity-rectified); further, that even the spatial distribution of the extracted field maxima and minima of the maps extracted at the times of maximal field power closely resembles the mean of all maps (when polarity-rectified) /21/. In short, a combination of the extractions yields a manageable and meaningful data set: First, the most pronounced maps at the times of maximal global field power are selected; then, the landscape pattern descriptors from these maps are extracted.

7. Adaptive Segmentation of Spontaneous EEG Map Series

Adaptive segmentation of the series of momentary maps during spontaneous EEG activity into epochs of stable spatial features has to consider the periodic polarity reversals of the EEG maps as a basic feature of living neuronal assemblies: the same landscape of inverted polarity is periodically generated. We accept a stable activity state to continue as long as the landscape stays constant, disregarding its polarity.

The locations of the maximum and minimum ("extreme") potential values in the maps at the times of maximal global field power are used for adaptive segmentation of the spontaneous map series into spatially stable segments, optionally after digital FIR filtering of the data to the temporal frequency band of interest /21/. A spatial window is erected around the locations of the extreme values of the first map. The window in our application included three neighbor electrodes. If in a subsequent map one or both extremes leave their spatial windows, the segment is terminated and a new segment starts with re-setting of the spatial windows. A segment is classed according to the two electrode locations where the extreme values were most frequently found.

A map of occurrences of extreme locations during the segment shows the main features as the more detailed map of the standard deviation of the mean of all maps during the segment (which corresponds to the Fourier transformation-derived power values of the waveshapes vs the average reference). Therefore, a more sensitive classification of the determined segments might be based on their power maps.

Segmentation of 2min recordings (16 channels) of spontaneous EEG during resting with closed eyes (8-12Hz band) in six normal subjects showed a mean segment duration of 210msec, a median duration of 250msec, with 50% of the recording time covered by segments longer than 323msec and 25% time covered by segments longer than 500msec /21/; the longest segment was 2560msec. An example of a sequence of segments is shown in Fig. 1. On the average over subjects, only 60 of the theoretically possible 120 segment classes were found, and the most prominent class covered 20% of total time, the two most prominent classes covered 32% of total time. The most prominent class showed right posterior and midline anterior extreme locations.

We conclude that on formal grounds there is evidence for the assumption that there is a sequence of brain functional states, concatenated by step-wise, non-continuous changes. The fact that 25% of the time was covered by segments longer than 500 msec is intriguing since Libet /22/ had shown that this is the minimal duration of afferent input at an intensity below which there is no conscious experience possible. Thus, space-oriented EEG time segmentation might be able to operationalize a putative time for conscious thoughts. Similar time spans as in our segmentation were reported for switching of attention (400msec: /23/), for effective times of leading visual masks (over 160msec: /24/), for backward masking (over 200msec: /25/) and for maximal inter-stimulus intervals for stereo percepts (250msec: W. Skrandies, pers. commun.). These observations suggest that early "components" in ERP map

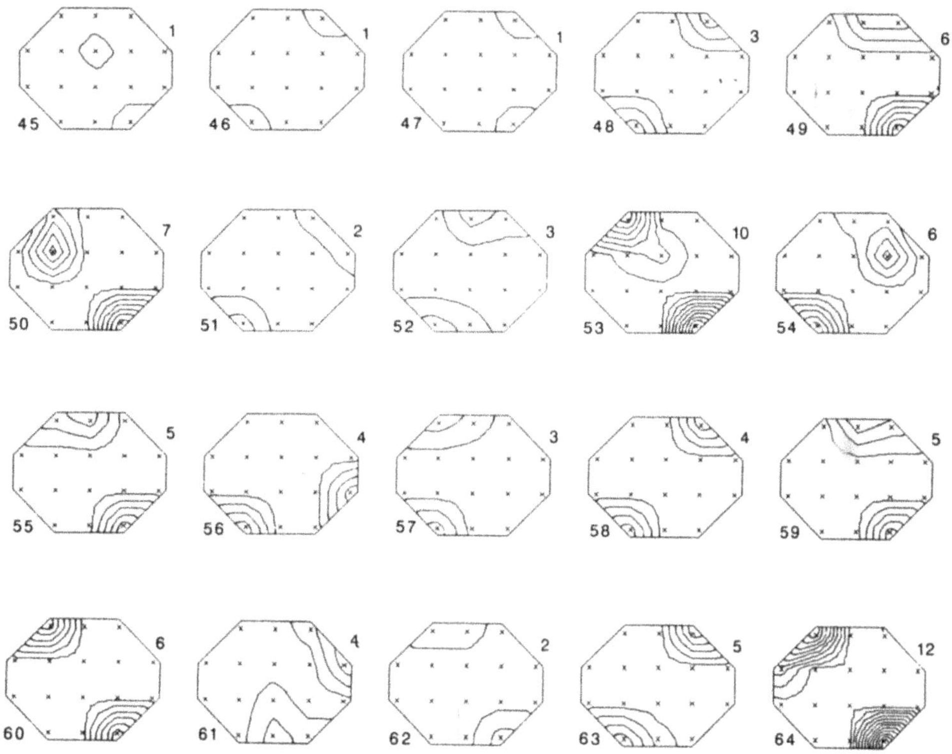

Fig. 1. Twenty successive segments (covering 4.7sec total time) of brain electric field maps recorded with 16 electrodes (dots) during a resting condition, with closed eyes, in an awake, normal subject. The segments were determined by the discussed procedure as epochs of near-stable map landscape. For each segment, the map shows the locations of occurrence of maximal and minimal potential values on the scalp between vertex and neck (head seen from above, nose up, left ear left). The occurrences were determined at time intervals of 50msec, and accordingly, the density of the equi-frequency-of-occurrence contour lines reflects segment duration; segment duration is also indicated by the number in the right upper corner wich is a count of the 50msec intervals during which the segment existed (this number multiplied by 50 therefore gives the segment duration). A running segment number is in the lower left corner. - Note the unequal durations, and the apparently irregular sequence of the different segments; note also that some segment classes occur several times (e.g. 53, 60, 62, 64), others only once (e.g. 54). From /21/.

series may not qualify for associated conscious experience of the ERP-identifiable processing steps because of their short durations.

There appeared to be no simple rule for the sequence of the different segment classes in our data, but comprehensive studies still need to be done (see also the paper by Fuchs et al in this volume). We do not, however, expect predictability, as we hypothesize that the different micro states are installed as a consequence of the preceeding mentation which is determined by the momentary state, the newly arriving information and the retrieved context recall from the accessible memory store, and hence, to a crucial extent reflects the individual-specific importance of the momentarily treated information /1/.

The inherent periodicity of the spontaneous EEG phenomena might be an organizing factor which provides a structuring of time slots during which different system states are implemented. The basic time structuring, manifest as peaks of global field power per unit time, changes for different gross states, in the range of seconds or minutes. Depending on the different time structuring, the functional significance of a momentary field distribution might be different. Within gross states, the momentary spatial configurations manifest the possible micro states. An appropriate procedure to analyze extended EEG map sequences accordingly might first adaptively segment the curve of global field power with conventional time-oriented procedures /e.g. 26/, in order to establish functional super-classes. In a second step, the spatially defined micro segments within the super classes are determined.

8. Functional Significance of Spontaneous Segments

In an attempt to identify the functional significance of different segment classes we studied motor reaction time to randomly occurring, infrequent high-pitch tone pips in a series of frequent lower-pitch tone pips, i.e. during the "oddball paradigm" of selective attention. Eight subjects had to respond to 337 rare tones each. The segment class immediately before each rare tone and the following reaction time was determined. Over subjects, reaction times were significantly different between segment classes before stimulus arrival (ANOVA p smaller than .0001), and there were similar differences in reaction time for the 26 different segment classes which occurred in all subjects (FRIEDMAN ANOVA chi square 40.6, df=25, p smaller than .025). This shows that brain functional micro state during wakefulness, as diagnosed by one instantaneous time slice of the EEG before information arrival, measurably influences simple behavior results of information processing.

At present we are conducting a pilot study on possible relations between spontaneous, momentary segment classes and cognition modes: Upon a tone signal, the subject reports the ongoing thoughts (this experimental design is similar to that of the studies on mentation discussed in "characteristics of states"). The reports are classed (e.g. abstract vs visualized thought), and compared to the segment class immediately prior to the signal. The present results have shown significant differences between the segment classes associated with the two cognition classes.

9. Segmentation of Series of Event-Related Brain Field Maps

The concept of segmentation of brain electric field map series into spatially stable segments can also be applied to event-related potential (ERP) field map series /9,14/. In this case, there is no need to consider periodic reoccurrences of maps of inverted polarity, since it is generally accepted that ERP "components" are aperiodic during the first about 500msec post-stimulus. (However, there might be an iterative time-structuring process underlying the apparent non-periodicity, as suggested by examining the global similarity of consecutive pairs of ERP maps /27/). If there is no periodic process, the polarity of the extreme values plays a decisive role in the classification of potential distribution landscapes, i.e. of momentary functional staes. Accordingly, in our segmentation procedure, an ERP segment (which represents a given functional micro state) terminates as soon as either the maximum or the minimum leaves its spatial window.

In a practical application of this adaptive segmentation procedure one determines the grand mean locations of maxima and minima per map over all experimental conditions in each subject, averages the locations of the maxima and minima over subjects, and thus obtains mean locations which are not restricted to measured electrode locations. This data set then is segmented, using spatial windows of, say, +/- 1/2 electrode distance around the location of the maximum and the minimum of the first maps of the series. When in a subsequent map one or both windows are violated, a new segment is started.

Segmentations of event-related map series of 400 to 800 msec duration after the presentation of the information to the subject (using spatial windows of +/- 1/2 electrode distance and the polarity-considering rule), yielded 5 to 12 segments. These segments were obviously much shorter than those found during spontaneous activity with the same window but without the polarity-considering rule. The event-related segments are viewed as the manifestations of different processing steps of the information which was received by the brain. For further comparisons between experimental conditions, the maps during each segment are averaged for each experimental condition, e.g. for "attention to the visual target" and "no attention to the visual target". The landscapes of these mean maps can be tested for significant differences between experimental conditions. In this way, similarities and dissimilarities between the electric activity patterns can suggest similarities and dissimilarities between brain mechanisms which are active in different perceptual or task conditions, and the time of occurrence of such differences after information input to the brain can be established.

In a study of the effects of voluntary attention and of subjectively visible contour figures ("Kanizsa figures", see also paper by Baumgartner et al in this volume) on event-related potential field map series we observed similar effects on map landscape in two different experimental conditions: (1) when voluntary attention was directed to a "non-figure" target, and (2) when a subjectively visible contour figure target was viewed by the subjects. The effects became first evident in the event-related potential field segment 168-200msec after the target presentation (see also /9/), and were measurable as increased anterior-posterior distance between the extreme potential values in the maps in comparison with maps obtained without attention, and without a contour figure. The effects persisted in some later segments (see Fig. 2). This finding is in agreement with the hypothesis that figure perception utilizes a brain resource which is also activated during voluntary attention, and that brain mechanisms of attention do not simply recruit more elements of the neural population which is engaged in the processing of a given sensory

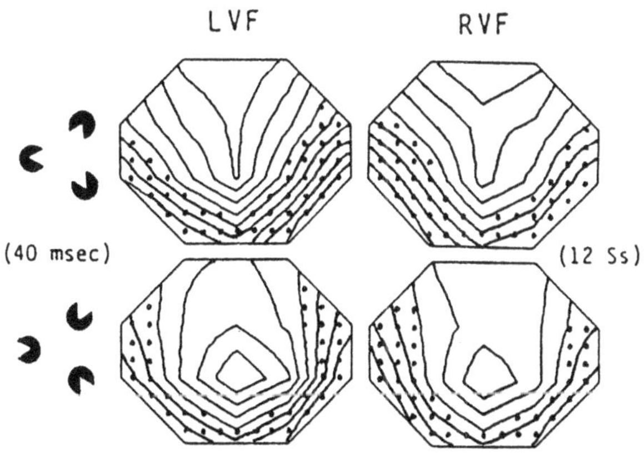

Fig. 2. Brain electric field (ERP) maps, mean maps during a time segment between 200 and 232msec after 40msec presentations of a subjective visual contour (KANIZSA) figure (upper maps) and a control figure without subjective contour (lower maps) in the left (LVF) or right (RVF) visual hemifield. - Head seen from above, nose up, left ear left. Recordings with 16 scalp electrodes between vertex and inion. Dotted negative, white positive, equipotential lines in .5 microvolt steps. Mean over 12 normal subjects, 160 stimuli/subject (from D. Brandeis and D. Lehmann, in preparation).

information, but that an at least partially different neural population becomes activated during a certain brief epoch starting 168msec after information arrival. - Other event-related potential field map studies of higher brain functions such as letter recognition and language understanding are discussed in /9/.

10. Review

The continuous stream of functional states of the brain is manifest as quickly changing different "landscapes" of the brain electric field which is recorded on the human scalp as a series of momentary maps of potential distributions. This continuum can adaptively be segmented into time epochs of unequal durations in the sub-second-to-second range during which the electric landscape is quasi stable, i.e. during which a homogenous neural process might be assumed to be active. These different brain micro states thus recognized can be shown to be associated with modes or steps of processing of incoming information, which measurably influence the behavioral outcome of information processing.

Acknowledgement
The research work in our laboratory has been supported in part by the Swiss National Science Foundation, the Hartmann-Müller Foundation (Zürich), the EMDO Foundation (Zürich), and the Roche Research Foundation (Basel).

References

1. Koukkou, M. and D. Lehmann: Brit. J. Psychiat. 142: 221-231 (1983)
2. Woodworth, R.S. and H. Schlosberg: Experimental Psychology, (Holt, New York, 1954)
3. Ehrlichman, H. and M.S. Wiener: Psychophysiol., 17: 228-235 (1980)
4. Lehmann, D. and M. Koukkou: Electroenceph. Clin. Neurophysiol., 37: 73-84 (1974)
5. Lehmann, D., M. Koukkou and A. Andreae: Sleep Res., 10: 152 (1981)
6. Berkhout, J., D.O. Walter and W.R. Adey: Electroenceph. clin. Neurophysiol., 27: 457-469 (1969)
7. Katada, A., H. Ozaki, H. Suzuki and K. Suhara: Electroenceph. Clin. Neurophysiol., 52: 192-201 (1981)
8. Rohrbaugh, J.W. In Varieties of Attention ed. by R. Parasuraman and D.R. Davies (Academic Press, New York, 1984)
9. Brandeis, D. and D. Lehmann: Neuropsychologia, 24: 151-168 (1986)
10. Borbely, A.A., P. Mattmann, M. Loepfe, I. Fellmann, M. Gerne, I. Strauch and D. Lehmann: Eur. J. Pharmacol. 89:157-161 (1983)
11. Lehmann, D., G.W. Beeler and D.H. Fender: Electroenceph. clin. Neurophysiol., 1965, 19: 336-343 (1965)
12. Keesey, U.T. and D.J. Nichols: Vision Res., 7: 859-879 (1967)
13. Williams, H.L., H.C. Morlock and J.J. Morlock: Psychophysiol., 2: 208-215 (1966)
14. Lehmann, D. and W. Skrandies: Progr. Neurobiol., 23: 227-250 (1984)
15. Lehmann, D.: In Synchronization of EEG Activity in Epilepsies ed. by H.H. Petsche and M.A.B. Brazier (Springer, Wien, 1972) p. 307-326
16. Lehmann, D.: In Computerized EEG Analysis - CEAN ed. by E. Dolce and H. Künkel (Fischer, Stuttgart, 1975) p. 102-110.
17. Lehmann, D.: Int. J. Psychophysiol., 1:267-276 (1984)
18. Lehmann, D.: Electroenceph. Clin. Neurophysiol., 31: 439-449 (1971)
19. Lehmann, D. and W. Skrandies: Electroenceph. clin. Neurophysiol., 48: 609-621 (1980)
20. Lehmann, D.: In Recent Advances in EEG and EMG Data Processing ed. by N. Yamaguchi and K. Fujisawa (Elsevier, Amsterdam 1981) p. 117-132

21. Lehmann, D., H. Ozaki and I. Pal: Electroenceph. Clin. Neurophysiol., in press (1987)
22. Libet, B.: Human Neurobiol., 1: 235-242 (1982)
23. Reeves, A. and G. Sperling: Psychol. Rev., 93: 180-206 (1986)
24. DiLollo, V.: J. exp. Psychol. General, 109: 75-97 (1980)
25. Michaels, C.F. and M.T. Turvey: Psychol. Res., 41: 1-61 (1979)
26. Creutzfeldt, O.D., G. Bodenstein and J.S. Barlow: Electroenceph. clin. Neurophysiol., 1985, 60: 373-393 (1985)
27. Lehmann, D.: In *Evoked Potentials III* ed. by C. Barber and T. Blum (Butterworths, London, in press 1987)

Spatio-Temporal Analysis of Multi-channel Alpha EEG Map Series

A. Fuchs[1], R. Friedrich[1], H. Haken[1], and D. Lehmann[2]

[1]Institut für theoretische Physik und Synergetik, Universität Stuttgart, Pfaffenwaldring 57, D-7000 Stuttgart 80, Fed. Rep. of Germany

[2]Neurologie, Universitätsspital Zürich, CH-8091 Zürich, Switzerland

1. Introduction

In this paper we wish to analyze a multichannel alpha EEG recording from a normal person. In these EEG measurements, time series (Fig. 1) were derived from various locations at the skull as schematically shown in Fig. 2.

In the first step of our analysis we interpolate the spatial dependence so that we are dealing with a continuous field of EEG patterns which is, however, time dependent. In the following we wish to analyze these data by considering the brain as a complex system whose activity undergoes a continuous change in time. We shall perform this analysis along two entirely different lines.

1. Along the lines of synergetics, which have been outlined in the contribution by H.Haken to these proceedings and which we shall discuss briefly below.
2. In a purely mathematical fashion by the Loève-Karhunen expansion which allows one to make the best guesses with respect to feature selection in pattern recognition.

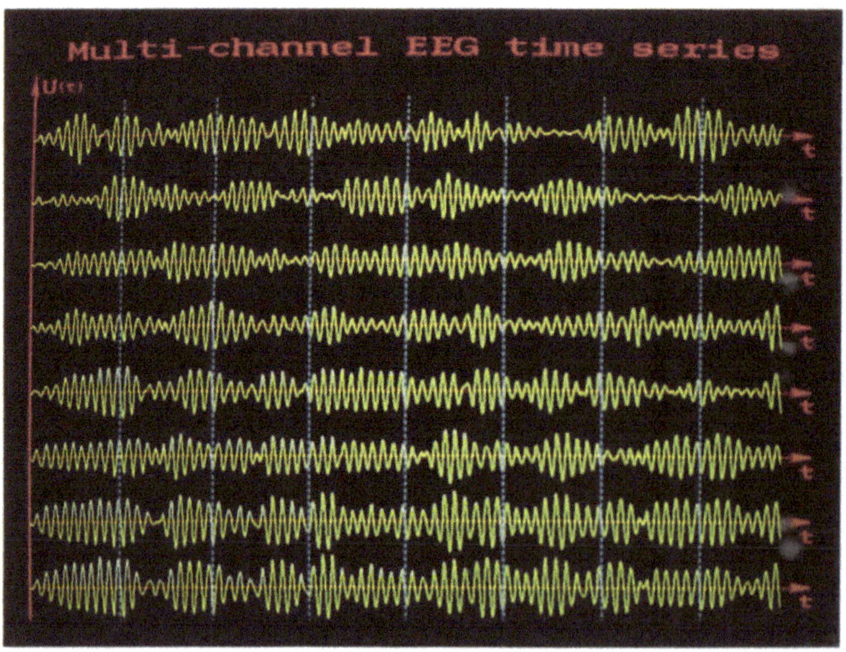

Fig. 1

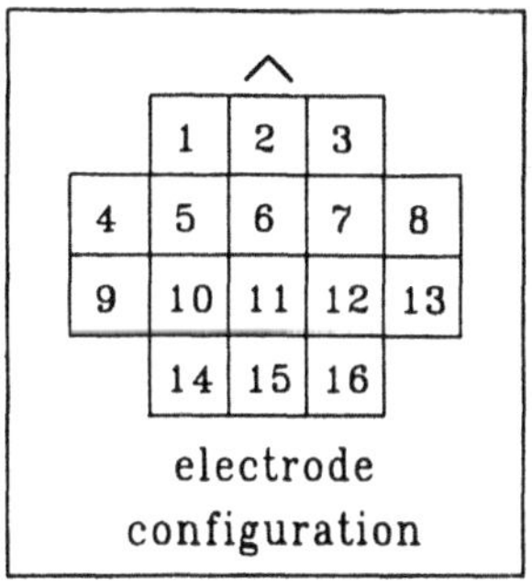

Fig. 2

Let us now comment on the first approach in some detail. The results of the application of the Loève-Karhunen expansion will be presented in Sect. 4.

In synergetics [1,2], it is by now well established that complex systems show an amazingly similar kind of behavior when they undergo dramatic changes of their spatial, temporal or functional structures at macroscopic scales. When such structures are formed, in general the dynamics is governed by very few degrees of freedom only. These degrees of freedom are usually connected with different macroscopic spatial patterns. A typical example may be quoted from the paper of Friedrich and Haken in a previous volume of the Springer Series in Synergetics [3]. Here a model of a planet's atmosphere was examined and it was shown that specific convection patterns may occur which are superpositions of few basic patterns. A warning should be added at this moment immediately. This analogy is purely formal and we do not claim by any means that the processes going on in the brain have anything to do with, say, convection in liquids. So the analogies which are established here are made at a very high abstract level which is of purely mathematical nature. But the essence remains the same, namely in situations where patterns are formed by means of self-organization just a few degrees of freedom may dominate the dynamics. Then the total pattern is composed of a superposition of the individual 'modes'. These modes describe the spatial behaviour belonging to the different degrees of freedom which determine the time dependence, and are denoted as order parameters in synergetics.

Let us now try to establish an analogy between the systematic treatment of spontaneous pattern formation in a physical system, say a system in hydrodynamics, and the phenomenological approach to the analysis of spatio-temporal patterns occurring in a complex system represented, in the present case, by the EEG maps (see Table 1).

On the microscopic level physical systems are composed of atoms or molecules, whose number is of about 10^{22} in a cubic centimeter. Their behavior is governed by the equations of motion derived from the laws of nature in physics. In the corresponding case of neurophysiology we have to deal with neurons, say of a number of 10^{11} and they may be characterized for instance by their firing activity. However, very little is known about the basic laws governing the neural behaviour, except for some important but still insufficient knowledge about physiological data of neurons.

At the mesoscopic level we lump the fluid particles together to volume elements where we define temperature, velocity, and density fields. For these quantities it is possible to establish evolution equations (cf. the contribution by H. Haken). The question which immediately arises in the analogous case is how to describe the behaviour of ensembles of neurons adequately. Our results may suggest that the neural activity as manifested in the corresponding EEG patterns can be characterized by a more or less continuously changing field of the electric potential. But great care must be exercised with respect to this interpretation because the field of the neurons may be changed considerably by the dielectric properties of the skull so that we may have a smearing out of small scale structures of this field when measured from the outside of the skull. We shall not dwell on this problem Table 1 here but rather consider the EEG data as phenomenologically given. Then we wish to draw all the conclusions from these given data which are possible in the present context.

At the macroscopic level then, it is well known from synergetics that specific structures may occur in a self-organized fashion where the individual parts of the system show collective behaviour giving rise to spatio-temporal patterns. As mentioned above, these patterns are governed by few order parameters which, in turn, obey the so-called generalized Ginzburg-Landau equations. The question which is exhibited in Table 1 is now, how to draw conclusions on the mesoscopic level in the case of the theoretical explanation of neural activity. Here we have time scales of 0.1 sec and length scales of the size of the human brain. As shown in Figs. 3 and 4 we find coherent spatial patterns with complex temporal dynamics. The basic idea will be: what are the order parameters and what are the equations describing their temporal evolutions. In the present paper we shall be able to identify the order parameters. An analysis of their dynamics is presently being performed and will be published elsewhere.

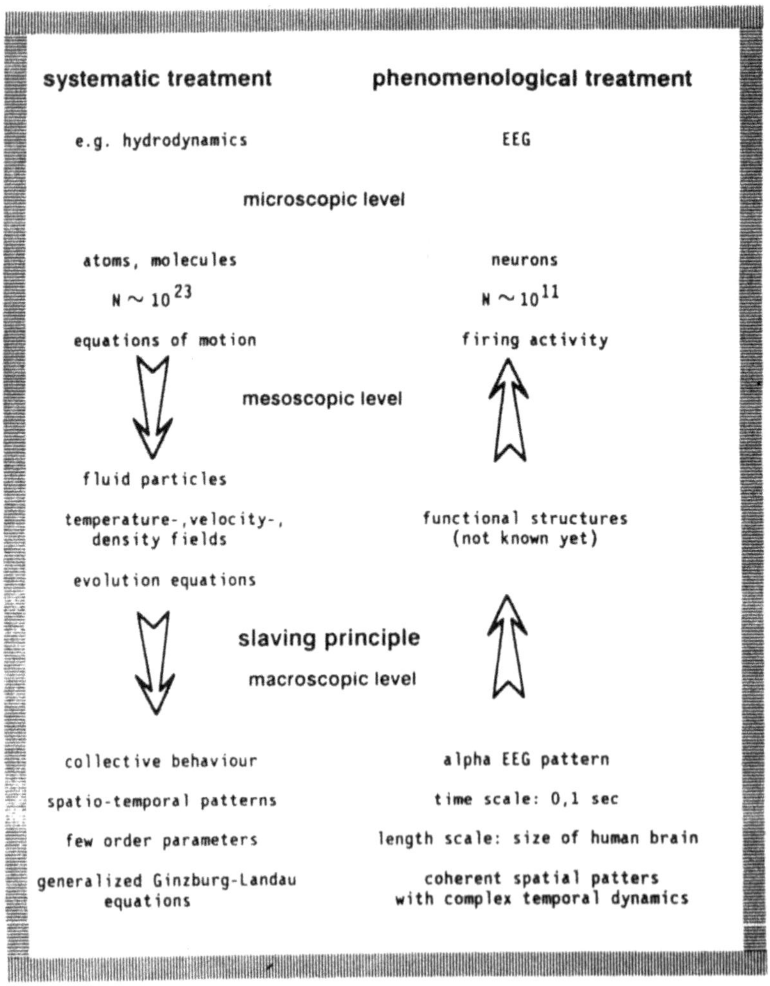

Table 1

In the analysis used by Lehmann et al. (this volume, see also [4]) the different classes of spatial distribution of the measured potentials are emphasized, and are viewed as series of very brief individual brain states. The question of state sequence was suggested to be controlled by incoming or retrieved information. The present paper, however, comes to the conclusion that in the case of alpha-activity it should be possible to view these state sequences as a time evolution governed by deterministic laws.

We shall proceed as follows. In Sect. 2 the experimental set-up and the process of data acquisition will be described. Section 3 is devoted to the analysis of the obtained alpha EEG series, while Sect. 4 will present the Loève-Karhunen analysis.

2. Experimental Set-Up

The EEG data were recorded in 16 channels from a normal unselected, right-handed, male volunteer of 35 years of age. Grass gold cup electrodes were attached with Grass EC-2 cream at about equidistant spacing over an area between the vertex and 0.5 cm above the inion. The array scheme is shown in Fig. 2. The electrode at the vertex was used as a common reference. The spontaneous EEG was recorded from the comfortably seated subject in a sound-shielded chamber. After 5 to 8 minutes of adaption to the recording surroundings, data collection started 30 s after the request to close the eyes. After amplification (GRASS P5II preamps) and anti-alias filtering, the data were sampled for 2 minutes at 128 samples/s. Digital filtering was done off-line, using a finite impulse-response linear phase filter with a band pass of 8.6-11.6 Hz (3dB points; 20 dB down at 7 and 13 Hz).

3. A detailed analysis of alpha EEG map series

The obtained data are originally registered as time series of the electric potential of the individual points of the skull. The question to be checked first is whether the individual time series are connected with spatial patterns which change on a larger length scale as compared to the distance of the different grid-points. To this end we first interpolated between the spatial points and prepared a movie which clearly showed the well-known spatial coherence of the patterns. A first inspection reveals that there seem to be two main characteristics of the dynamics: A rotation of the pattern which jumps more or less irregularly from the left to the right direction and vice versa as discussed by Lehmann [5]. This rotation is superposed by an additional oscillation of the field strength of the electric potential (Fig. 3). The second sequence (Fig. 4) consists in a, roughly speaking, nonrotating, standing pattern, whose activity undergoes temporal oscillations.

In order to cast these observations into a mathematical form we first have to formulate a basal surface with associated boundary conditions. Then using the analogy with the many other problems treated in synergetics, e.g. the paper by Friedrich and Haken mentioned above, we shall attempt to decompose the patterns into a few basic modes. Let us discuss these problems in detail.

1. Choice of geometry

   The rotations and oscillations are compatible with a circular area. Therefore we shall make a map as exhibited in Fig. 5.

2. Boundary conditions

   The boundary conditions of the patterns are unknown and probably not well defined. To cast the problem into a mathematical form we embed the observed area into a larger area (compare Fig. 6). Here we have chosen the boundary condition such that the amplitude of the pattern vanishes at the border

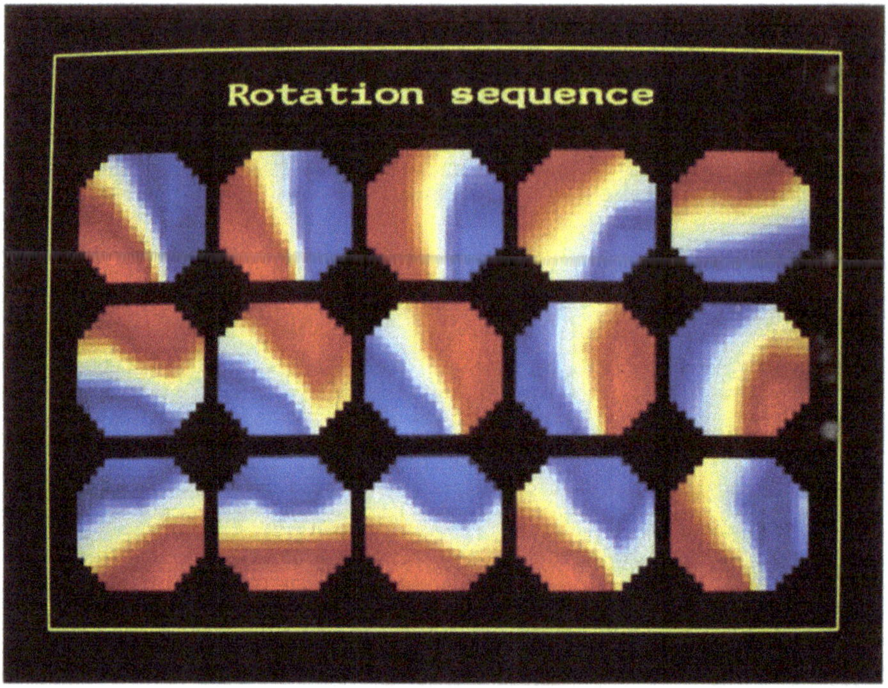

Fig. 3

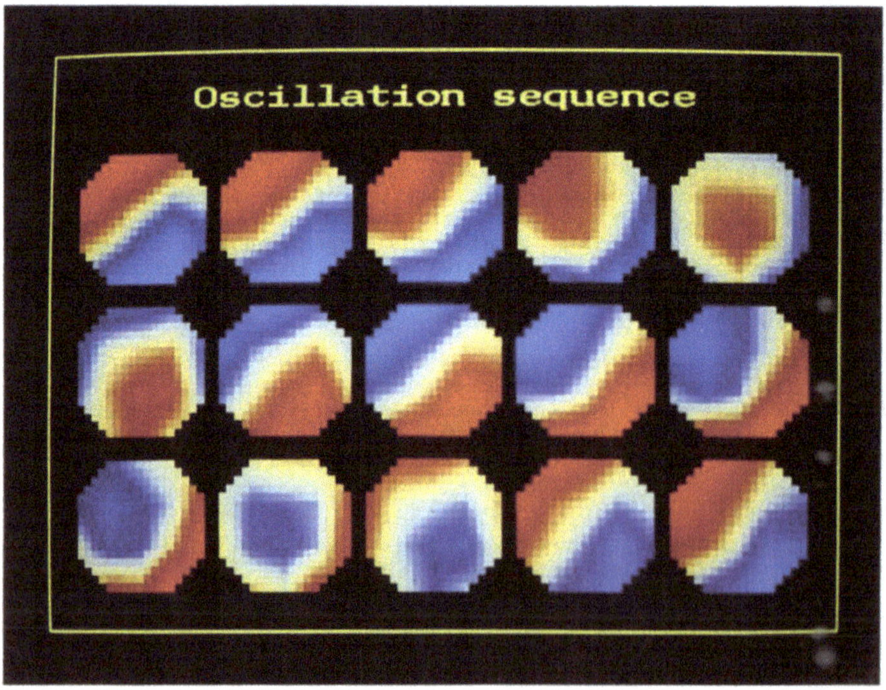

Fig. 4

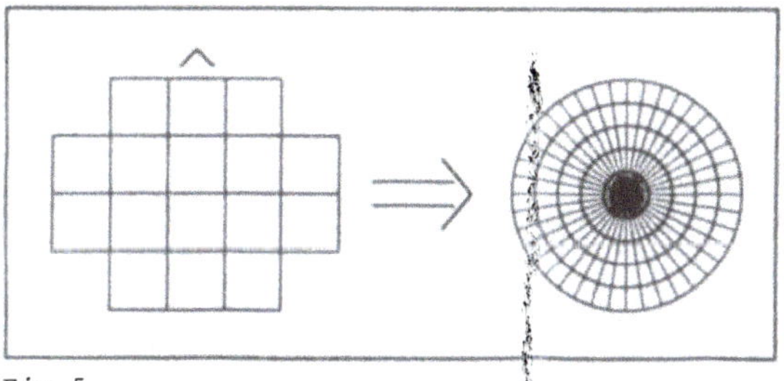

Fig.5

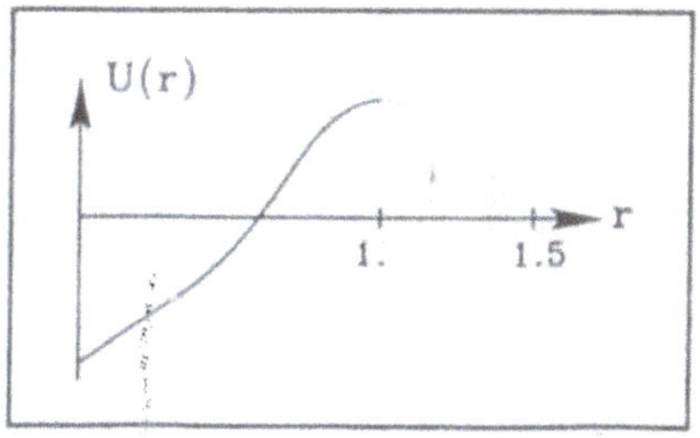

Fig.6

line. The typical radial dependence of the pattern is shown in fig. 6 where $r = 1$ is the actual boundary whereas the value $r = 1.5$ is chosen so that here the pattern may be assumed to vanish.

Now we are able to perform a mode expansion of the patterns. In view of the circular geometry and having in mind typical modes occurring in synergetics we are led to consider modes which, in the azimuthal direction are exponential functions of the angle ϕ and, in the radial direction, Bessel functions of the first kind. This two-dimensional basis is shown in Fig. 7.

The decomposition of the the pattern $U(r,\phi,t)$, which was derived from the experimental data by interpolation is done by the following expansion:

$$U(r,\phi,t) = \sum_{k,m} \xi_{km}(t)\, J_m^k(r)\, e^{im\phi} \qquad (1)$$

where $\int_0^R r\, dr\, J_m^k(r)\, J_m^{k'}(r) = \delta_{k,k'}$.

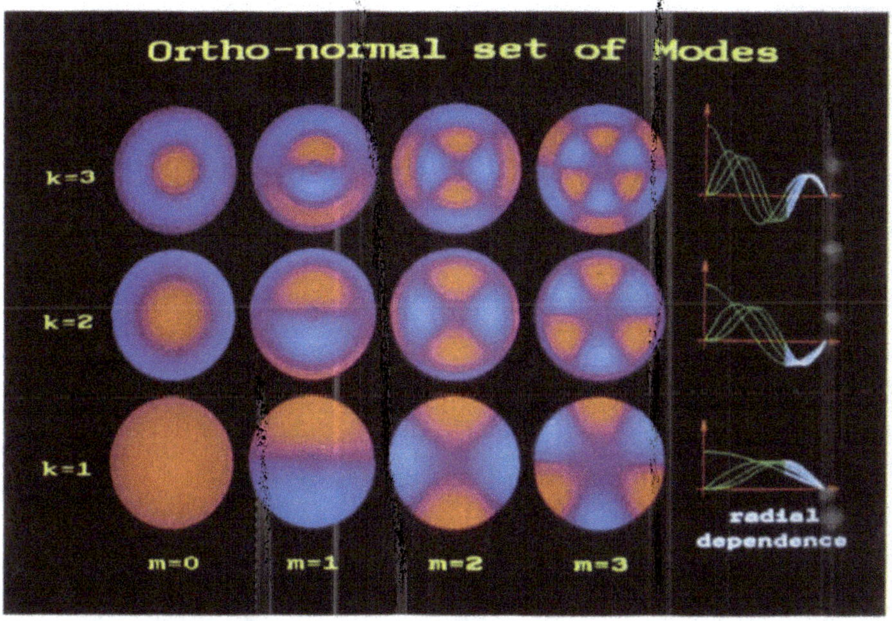

Fig. 7

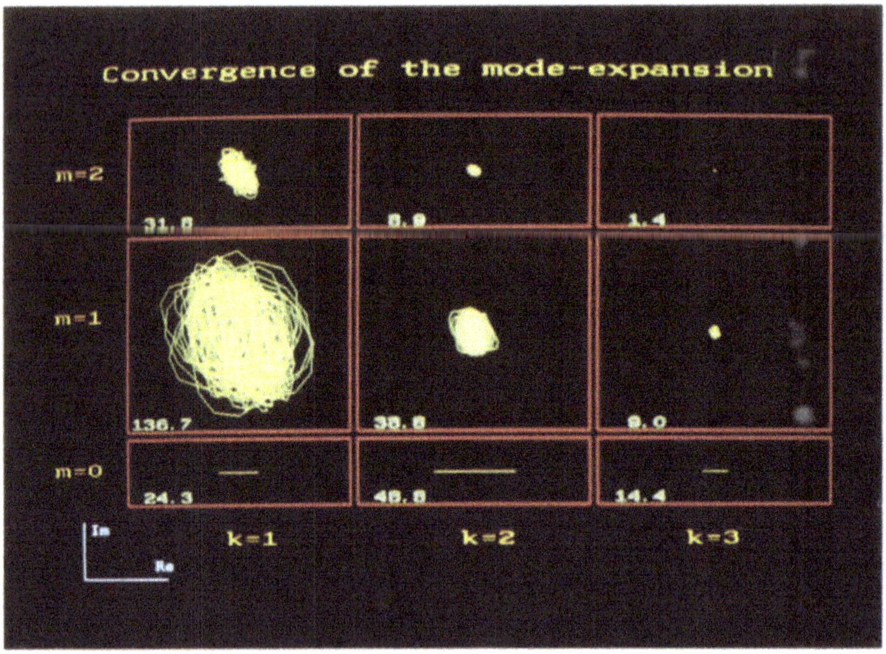
Convergence of the mode-expansion
m=2
m=1
m=0
1.4
136.7
24.3
14.4
Im
Re
k=1
k=2
k=3

Fig. 8

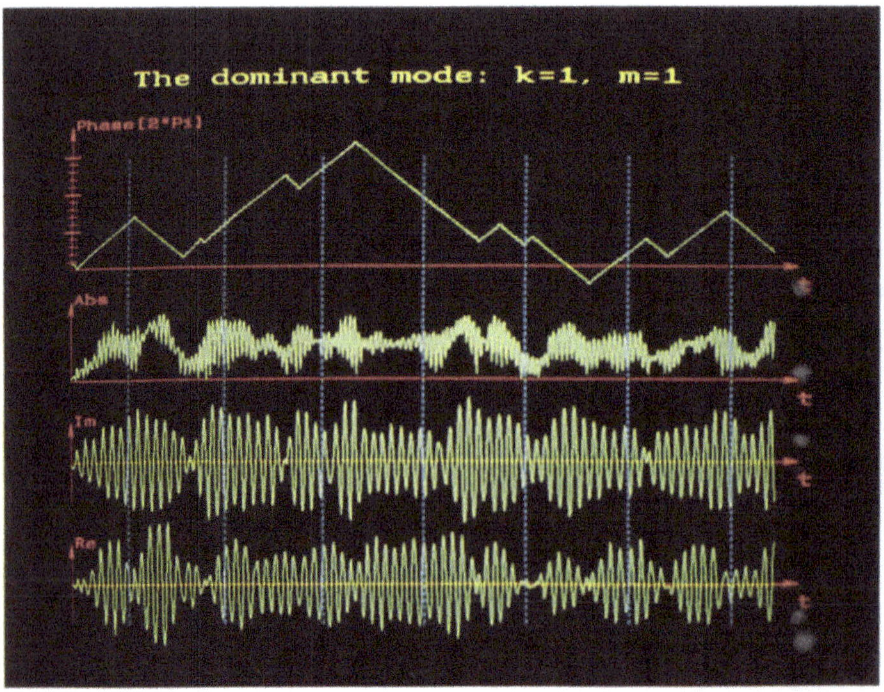
The dominant mode: k=1, m=1
Phase[2*Pi]
Abs
Im
Re
t

Fig. 9

The expansion coefficients $\xi_{km}(t)$ are thus given by

$$\xi_{km}(t) = \frac{1}{2\pi} \int_0^{2\pi} d\phi \int_0^R r \, dr \, J_m^k(r) \, e^{-im\phi} \, U(r,\phi,t) \tag{2}$$

The evaluation of the integrals in (2) was done numerically. We obtain the following results:

1. The expansion converges with increasing k and m (see Fig. 8).
2. $\xi_{11}(t)$ is the most dominant mode.
3. $\xi_{km}(t) = \xi_{km}(t) \, e^{i\phi(t)}$, where $\phi = \pm \omega$ is a slowly varying quantity (see Fig. 9).

From 2. and 3. we can derive the main characteristics of the dynamics. We find the left-right rotation and the oscillating sequences of the standing patterns mainly governed by modes with $m = \pm 1$.

In the next step we utilize these results to make a separation of the fast varying rotation and the slowly varying amplitudes according to

$$\xi_{km}(t) = \eta_{km}^{\ell}(t) \, e^{i\omega t} + \eta_{km}^{r}(t) \, e^{-i\omega t} \tag{3}$$

where $\omega = 63/s$ which corresponds to the frequency of 10/s in the alpha-EEG time series. The amplitudes $\eta_{km}^{\ell,r}(t)$ are given by

$$\eta_{km}^{\ell,r}(t) = \frac{1}{T} \int_{t-T/2}^{t+T/2} dt' \, e^{\pm i\omega t'} \, \xi_{km}(t') \quad . \tag{4}$$

The validity of this rotating wave approximation has been numerically justified by calculating the error

$$\varepsilon = \xi_{km}(t) - \eta_{km}^{\ell}(t) \, e^{i\omega t} - \eta_{km}^{r}(t) \, e^{-i\omega t} \tag{5}$$

which turns out to be a small quantity. If we plot the slowly varying modes as functions of time (Fig. 10) we immediately realize that modes with different k and the same m are strongly correlated. This leads us to introduce new modes $f_m(r)$ in the radial direction:

$$f_m(r) = \sum_k a_k^m \, J_m^k(r) \quad . \tag{6}$$

As a result, our analysis shows that we can reconstruct the experimentally observed patterns by a superposition of only three independent spatial modes. These modes belong to $m = 0,1,2$. The mode amplitudes are complex, while the one belonging to $m = 0$ is real. Therefore we have 5 independent variables to describe the spatial pattern at each time step.

Some care must be exercised with respect to the interpretation of our results. They show that there are only three spatial modes underlying the whole analysis. On the other hand the coefficients of these spatial modes, which are time dependent, belong to a dynamical system whose dimension is not yet elaborated on. Future work will be devoted to determining Lyapunov exponents and the dimensions of the corresponding processes.

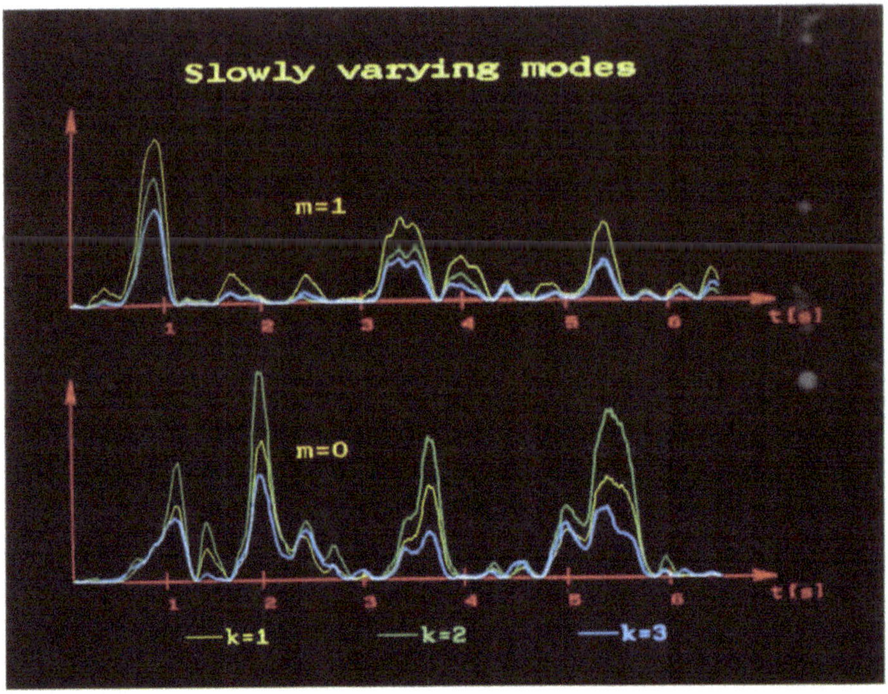
Slowly varying modes
m=1
t[s]
m=0
1
2
3
4
5
6
t[s]
—k=1
—k=2
—k=3

Fig. 10

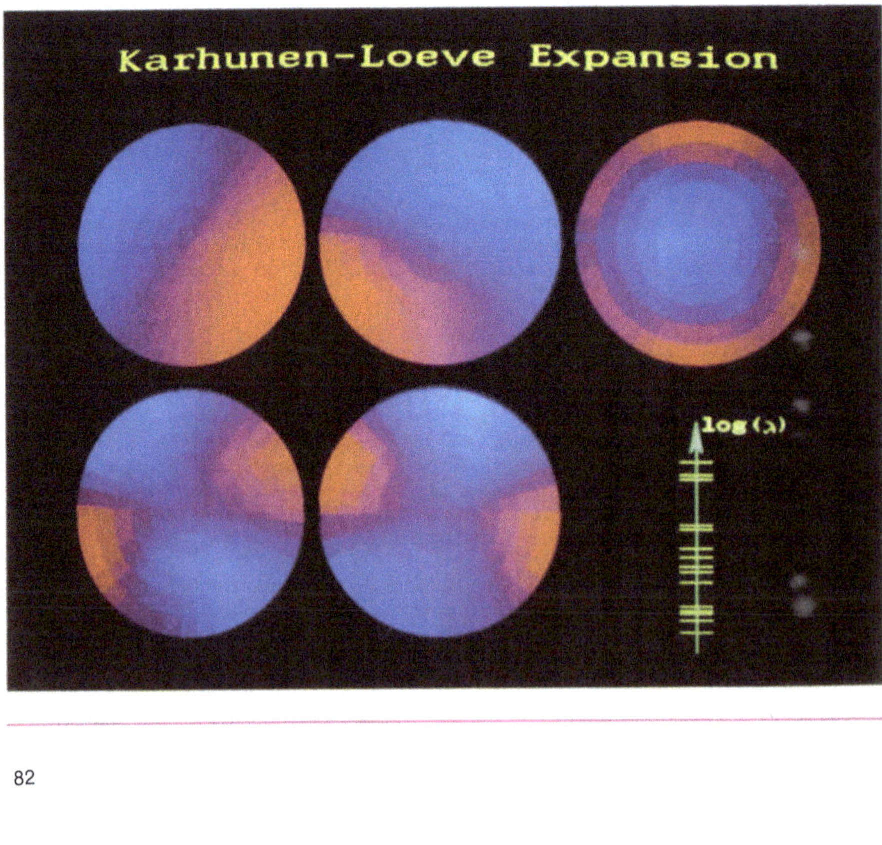
Karhunen-Loeve Expansion
log(λ)

Fig. 11

Let us summarize the method we have employed. We used as a guideline general ideas of synergetics on pattern formation in complex systems. We have shown that we could easily identify the underlying mode patterns and that, in reconstructing the patterns by a superposition of the order parameters, we could find excellent agreement with the experimentally given data. In the next section we shall analyze the patterns using the Loève-Karhunen expansion of pattern recognition theory.

4. Loève-Karhunen Expansion

As is well known in pattern recognition theory [6], the Loève-Karhunen expansion allows one to extract from given patterns those features which are the most significant ones, i.e. which contain the greatest amount of information. To this end one considers the correlation matrix averaged over the whole time series and diagonalizes it. The eigenvectors are then the features while the eigenvalues describe the dominance of the various features in the whole decomposition. To this end we calculated the averaged correlation matrix for the 16 points given, so that we had to diagonalize a 16 by 16 matrix.

The results are presented in Fig. 11. It turns out that the spatial dependences of the eigenvectors which correspond to the 5 largest eigenvalues are of the same type as the basic modes we used in Sect. 3. This agreement is a further argument in view of the interpretation of our results by basic ideas of synergetics.

Acknowledgements

We thank M. Bestehorn and K. Marx for valuable discussions and acknowledge the aid of W. Lorenz who took the colored photographs.

References

[1] H. Haken, Synergetics, An Introduction (Springer 1978)

[2] H. Haken, Advanced Synergetics (Springer 1983)

[3] R. Friedrich, H. Haken in Complex Systems, Operational Approaches, Ed. H. Haken (Springer 1985)

[4] D. Lehmann, H. Ozaki, I. Pal EEG alpha map series: Micro states by space-oriented adaptive segmention Electroenceph. clin. Neurophysiol. (1987) in press.

[5] D. Lehmann, Multichannel topography of human alpha EEG fields. Electroenceph. clin. Neurophysiol. 31. 439-449 (1971)

[6] S. Watanabe, Pattern Recognition: Human and Mechanical (John Wiley & Sons 1985)

Part III

Pattern Analysis and Pattern Formation

Singular System Analysis of Time Series Data

E.R. Pike

King's College, London, United Kingdom and
Royal Signals and Radar Establishment, St. Andrews Road,
Malvern, Worcestershire, WR14 3PS, United Kingdom

1. Introduction

In this contribution I will review work by myself, M Bertero of University of Genoa, C De Mol of the University of Brussels and colleagues at RSRE including D S Broomhead, G P King and R Jones.

The origin of the work has been the mathematical development of the theory of "Singular Systems" applied to the problem of inversion of the finite Laplace transform, following earlier work by McWhirter and Pike [1] in 1978 on the eigenfunction inversion of the infinite Laplace transform. The eigenfunction spectrum is a well-known concept in many areas of theoretical physics but its generalisation, in the form of the singular value spectrum, although well known in mathematics for more than half a century, has not enjoyed a great deal of application in practical science, although, Karhunen-Loeve transformations in statistics use closely related techniques as do recent radar algorithms such as MUSIC. In the next section we review inversion of first kind Fredholm equations by eigenvalue methods which give rise to the classical theory of information. In section 3 situations for which the eigenvalue method cannot be applied will be introduced and the method will be generalised for these other inversion problems. The recent work of Bertero et al, developing the theory of resolution limits for various non-invertible mappings will be reviewed. The beginnings of the application to time series data appear in the application of the singular system theory to truncated and sampled data and this will be discussed in section 4. We then consider general data vectors which may represent time series and follow Takens in constructing linear maps using windowed sequences. Takens initiative [2], however, is taken much further here and a number of the initial difficulties are circumvented in this new approach. Recent work of Broomhead, Jones and King [3] on the extraction of local dimension of phase space manifolds and recent work by Bertero et al [4], on an exact analytic inversion of an infinite matrix arising in scanning with a band-limited window function, are reviewed in the final section.

The presentation follows closely that given in Broomhead et al [5], but with the addition of the work on band-limited windowing.

2. Band-Limited Systems

The amount of information that we demand from finite data was discussed in the theory of band-limited communication in the early papers of Shannon [6]. The information may be quantified in various ways as a number of degrees of freedom or as a resolution limit.

The Rayleigh resolution limit in diffraction-limited imaging is a classical example. The Shannon number here is the number of Rayleigh resolution elements covering the object. Slepian and Pollak [7] solved the band-limited transmission problem, which is the same problem for diffraction-limited imaging and communication theory, in a detailed analysis in the early sixties. They solved the problem by expansions in eigenfunctions and introduced the basis of prolate spheroidal functions by finding a commuting differential operator for the integral operator defined by the Fredholm equation to be explained below.

A full solution of the problem including the lost cut-off frequencies was found in the mathematical analysis. This is possible due to the fact that the reconstruction is of an entire function and thus partial knowledge can be utilised by analytic continuation to find the complete solution.

For illustration, we confine ourselves to one dimension and pose the problem of band-limited (diffraction-limited) imaging in the form of a Fredholm equation of the first kind. An "object" $f(x)$ gives rise, after transmission through a band-limiting apparatus which passes all Fourier frequencies in full up to a limit Ω and annihilates the rest, to an 'image' $g(y)$:

$$g(y) = \int_{-X/2}^{X/2} \frac{\sin\Omega(y-x)}{\pi(y-x)} f(x)\, dx \,. \tag{1}$$

Slepian and Pollak found the eigenfunctions, ϕ_n, and eigenvalues, λ_n, of this problem, which obey

$$\int_{-X/2}^{X/2} \frac{\sin\Omega(y-x)}{\pi(y-x)} \phi_n(x)\, dx = \lambda_n \phi_n(y) \quad , \quad n = 0, 1, 2 \ldots \tag{2}$$

or

$$K\phi_n = \lambda_n \phi_n \tag{3}$$

by using the fact that a commuting differential operator can be found for the integral operator, K, for $L^2(-X/2,X/2)$ into itself defined by

$$g(y) = (Kf)(y) = \int_{-X/2}^{X/2} \frac{\sin\Omega(y-x)}{\pi(y-x)} f(x)\, dx \,, \; \frac{-X}{2} \leqslant y \leqslant \frac{X}{2} \,. \tag{4}$$

The formal solution, then, for the reconstruction $f(x)$ is:

$$f(x) = \sum_{n=0}^{\infty} \frac{1}{\lambda_n} \left[\int_{-X/2}^{X/2} g(y)\, \phi_n(y)\, dy \right] \phi_n(x) \tag{5}$$

which, writing the scalar product in L^2 as $\langle *,* \rangle$, may be expressed in the form

$$f(x) = \sum_{n=0}^{\infty} \frac{\langle g,\phi_n \rangle}{\lambda_n} \phi_n(x) \,. \tag{6}$$

The ϕ_n are a complete orthonormal basis set in $L^2(-X/2,X/2)$ called prolate spheroidal functions. The eigenvalue spectrum λ_n has the well-known spectacular behaviour of holding up nearly to unity from $n = 0$ until $n = X\Omega/\pi = S$, the "Shannon number", when it drops almost to zero for all higher values of n.

The experimental consequence is that in a decomposition of f into its components ϕ the data will contain the projection of f in a subspace of $L^2(-X/2,X/2)$ spanned by ϕ_0 to ϕ_S without significant distortion or loss, but that the projection of f in the orthogonal complement to this subspace will be severely attenuated on transmission and, in practice, lost in inevitable physical noise. Analytically, of course, in the absence of noise the low amplitude of the higher eigenvalues has no consequence and a full inversion is formally available.

A description in terms of resolution is given by considering the reconstruction $S(x,x_0)$ up to the "lost" components of a δ-function object, say at x_0. Thus

$$\langle g,\phi_n\rangle = \langle Kf,\phi_n\rangle = \langle f,K^*\phi_n\rangle = \lambda_n\langle f,\phi_n\rangle = \lambda_n\phi_n(x_0) \tag{7}$$

and hence

$$S(x,x_0) = \sum_{n=0}^{S} \phi_n(x_0)\ \phi_n(x) \ . \tag{8}$$

This "impulse response function" may be computed in any given case and has the form of a central lobe whose width decreases to zero as $S\to\infty$, flanked by smaller sidelobes. Some measure of the width of the central lobe can be used as the resolution available.

3. Theory of Singular Systems

In many cases we are faced with the inversion of mappings from one space, say X, into another, say Y, whereas above the mapping was from L^2 into itself. For example, in the diffraction-limited imaging problem we may wish to record data outside the geometrical image. In this case our integral operator maps $X = L^2(-X/2,X/2)$ into $Y = L^2(-\infty,\infty)$. More generally data is truncated out of necessity at some limits, say [c,d], and the object defined over others, say [a,b] and we thus have a mapping from $L^2(a,b)$ into $L^2(c,d)$. These problems do not have solutions in terms of eigenfunction expansions since an eigenfunction loses its definition except in relation to a mapping of a space into itself.

Fortunately, the required extension of the theory of eigenfunction decomposition has existed since the turn of the century [8] in the form of bi-orthogonal relationship between bases in the two spaces (the so-called "shifted" eigenvalue problem). Thus suppose we have a linear bijective mapping $K: X\to Y$, and an adjoint operation K^* such that

$$\langle Kf,g\rangle_Y = \langle f,K^*g\rangle_X \tag{9}$$

where Dirac's notation for the Hermitian conjugate and dual spaces are again used and the scalar products in X and Y are denoted by $\langle *,*\rangle_X$ and $\langle *,*\rangle_Y$, respectively. We now have the following generalisation of equation (3),

$$\begin{aligned} K|u_n\rangle_X &= \alpha_n|v_n\rangle_Y \\ K^*|v_n\rangle_Y &= \alpha_n|u_n\rangle_X \end{aligned} \tag{10}$$

where the $|u_n\rangle_X$ and $|v_n\rangle_Y$ are complete orthonormal basis sets in X and Y, respectively and the α_n are the "singular values" of the decomposition.

By substituting the second equation of (10) into the first, and vice versa, we have

$$KK^*|v_n\rangle_Y = \alpha_n^2|v_n\rangle_Y$$

and

$$K^*K|u_n\rangle_X = \alpha_n^2|u_n\rangle_X \ . \tag{11}$$

Thus the α_n^2's are the eigenvalues of the operator $KK^*:Y \to Y$ and also of the operator $K^*K:X \to X$, with $|v_n\rangle_Y$ and $|u_n\rangle_X$ the respective eigenfunctions.

We call the triple of mapping, object space, and image space $\{K,X,Y\}$ a singular system, with natural singular value spectrum and singular bases $\{\alpha_n:u_n,v_n\}$.

The use of the singular system to invert a Fredholm equation of the first kind,

$$g(y) = \int_a^b K(x,y)\, f(x)\, dx \qquad c \leqslant y \leqslant d \tag{12}$$

is a simple generalisation of the solution (6) for the "square" problem of the previous section. We have thus

$$f(x) = \sum_{n=0}^{S} \frac{\langle g,v_n\rangle_Y}{\alpha_n}\, u_n(x) \tag{13}$$

where the series expansion in X in the basis functions u with coefficients given by scalar products in Y with the basis functions v is truncated (or may be rolled off more smoothly if desired) at a value of n which is a generalisation of the Shannon number. This new number of degrees of freedom is not defined quite so sharply as is possible with the eigenvalue spectrum of the prolate spheroidal functions, where the level of noise present hardly affected the issue. In fact, the concept of number of bits of information and the idea of an information theory as such rested on the remarkable properties of this spectrum as discussed above.and hence was formulated without explicit reference to noise levels. We now have to decide at what point the transmitted singular component will be lost in noise and this will depend on how the singular value spectrum falls off (which it must if K is compact) and on the actual noise levels present.

The most important properties of this method of data reduction lie in the fact that the information in the coefficients of the singular functions is optimally compressed. That is to say that if a series of a fixed number of components is to be used to approximate the function, the least error will ensue if the singular basis is used. In the pattern recognition field the "representational entropy" is said to be a minimum, or in rougher terms the "moment of inertia" of the spectrum about the origin is the least possible.

4. Sampling and Truncation

The nature of physical experimentation is such that data will be sampled with a finite resolution and truncated over some reasonable limits. In other words the output of an experiment will be a vector in an N-dimensional vector space rather than an L^2 function.

Again, fortunately, the singular system theory can be adapted so that K is a mapping from an L^2 space X into a vector space Y. On the other hand, if the input happens to be best described as a vector in a vector space X, the K may be a rectangular matrix mapping $X \to Y$. In the first case K^*K is a finite rank integral operator and KK^* a square matrix. In the second case both KK^* and K^*K are square matrices (of different dimensions).

The same inversion procedure is applied and (13) again gives the result with $\langle *,*\rangle_Y$ defined appropriately as a Euclidean scalar product and with the u_n either L^2 functions (represented in actual computer calculations on an arbitrary fine mesh) in the former case or suitable vectors in the latter case.

Many inversion problems have been treated by these methods in recent years [9] and the technique gives both an excellent match to digital experimental data and an exact rejection of out-of-band noise. Furthermore, the technique lends a comprehension of the nature of the ill-conditioning which would arise if the generalised Shannon number were taken at too high a value. We might hope that in a well planned experiment we are not asked to recover "invisible" singular components which have been transmitted so weakly (if the value of α_n is so small) that the information required is masked by noise in the data. If this is not the case, then a reconstrution can only be given of that part of the object which lies in the "visible" projection on the subspace spanned by u_0 to u_S where S is the generalised Shannon number.

5. Time Series Data

The application of the above theory to dynamical systems rests on the fact that a manifold M of dimension m, on which the phase space trajectory evolves, can be embedded in a linear vector space, Y, of sufficiently large dimension. An embedding is a diffeomorphism between the manifold M and an image manifold in Y. The foundation of this is a theorem by Whitney [10] who proved that a smooth (C^2) m-dimensional manifold which is compact and Hausdorff may be embedded in R^{2m+1}. The dimension 2m+1 is, generically, the minimum required to ensure the embedding. Note that the proof given makes no assumptions about the detailed structure of M, thus some m-manifolds can be projected into fewer than 2m+1 dimensions.

As an example consider a closed 1-manifold (a limit cycle, for instance). The theorem states that it needs to be embedded in R^3, but most projections of a circle in R^3 will give an embedding in a plane. However, general 1-manifolds in R^3 will be knotted. For these there is no projection that does not give a self-intersecting image in the plane.

In previous work Broomhead and King [11,12] have used the singular system analysis to construct an embedding of the global attractor. The approach in that work considers time series data x_i of (N+n) samples at equal time steps and constructed from it N-dimensional vectors in R^N (which is to be associated with the space X of the previous sections) with basis set e_i given by

$$\begin{aligned} |e_1\rangle &= (1,0,0 \;\ldots\ldots\; 0)^T \\ |e_2\rangle &= (0,1,0 \;\ldots\ldots\; 0)^T \\ &\;\vdots \\ |e_N\rangle &= (0,0,0 \;\ldots\ldots\; 1)^T\ . \end{aligned} \tag{14}$$

Note that a point in R^N

$$|g_i\rangle = (x_i,\ x_{i+1},\ \ldots\ldots\ x_{i+N-1})^T \tag{15}$$

effectively represents the entire time series.

Takens proved that a set of n consecutive values from the time series give a vector on the image manifold in R^n (which is to be associated with the space Y of the previous sections) provided that $n > 2m+1$. Thus the ith point of the trajectory in R^n is denoted by

$$|y_i\rangle = (x_i,\ x_{i+1},\ \ldots\ x_{i+n-1})^T\ . \tag{16}$$

To describe the evolution of the trajectory we write

$$\langle y_i| = \langle e_i|K \tag{17}$$

where K is an Nxn matrix

$$K = \begin{vmatrix} y_1^T \\ y_2^T \\ \cdot \\ \cdot \\ \cdot \\ y_N^T \end{vmatrix} . \tag{18}$$

This "trajectory" matrix has the further properties that

$$\langle G| = \langle g_i|K \tag{19}$$

where the kth component of G is the autocorrelation coefficient

$$\sum_i x_i x_{i+k-1}$$

and

$$K|f_i\rangle = |g_i\rangle \tag{20}$$

where $|f_i\rangle$ are the unit basis vectors in R^n. Equation (20) is of the form discussed in the previous sections and we may thus use the singular system theory to find the "information" content of the data by projecting it onto the singular vectors of the operator K. In this case the smaller nxn matrix K*K is obviously the best to diagonalise and is, as can quickly be verified using (19), the matrix of correlation coefficients averaged over n consecutive samples of the time series. This matrix is the Karhunen-Loeve transformation which is used in statistics to obtain statistically independent processes.

As before the number of singular values, α_n, above a given "noise floor" will give us a number of degrees of freedom of the problem, and each singular component will be orthogonal and statistically independent from the others. However, the number of degrees of freedom defined in this way is <u>not</u> an invariant of the embedding process. A much more useful invariant is the <u>dimension</u> of the manifold. In the next section we will apply singular system methods to calculate this number.

6. The Dimension of the Manifold

Topologically irrelevant deformations of the manifold can change the apparent number of degrees of freedom. Locally, however, an m-dimensional manifold can always be approximated by the linear vector space R^m. Thus we may carry the analysis further to obtain the local dimension of the manifold by mapping not the whole trajectory from R^N to R^n but only those parts of it which lie, in a sufficiently small neighbourhood of some point chosen on the manifold. This work has been found to be related to previous work by Fukunaga and Olsen [13] but for a randomly sampled manifold.

The idea is firstly to reduce the problem by projecting the trajectory onto the "deterministic" subspace by shedding the noise subspace in the usual way, although this stage is only for computational convenience. Secondly, to select a point on this trajectory and to construct a neighbourhood matrix K' by rejecting those vectors which lie outside a ball of given radius r centred at this point (using the Euclidean norm). The singular values of K' are now widely separated into a set corresponding to the topological dimension of the manifold and a set corresponding to the residual noise. The singular values in the first set scale linearly with r and hence may be identified easily. Indeed, a simple calculation shows that for a manifold of m dimensions the m singular values are all equal and given by

$$\alpha = \left[\frac{N_b}{m+2}\right]^{\frac{1}{2}} r \tag{21}$$

where N_b is the number of points used in the neighbourhood.

We illustrate the above method with an example taken from [5]; a 2-torus. We have generated incommensurate sinusoids and passed them through a cubic nonlinearity of the form $x + x^3$. The data were sampled at 100 times the largest fundamental frequency. The resultant time series (quantised to 2^{14} levels) are shown in Fig. 1. Classical discrete Fourier transformation (with Hanning weights) of this time series produces the complicated spectrum shown in Fig. 2. The quantisation noise floor can be seen.

The time series were windowed to produce sequences of vectors in R^{25}. The corresponding singular spectrum is shown in Fig. 3 where there are five singular values above a band-limited noise floor; the noise being dominated by the quantisation of the data. Also shown in the figure are the singular vectors associated with the six largest singular values. Phase portraits constructed by projecting K onto the second and third singular vectors (c_2 and c_3) are shown in Fig. 4.

In each case when performing the local analysis, we assume the deterministic subspace to be spanned by the first five singular vectors. The reference points used in the calculations are those shown in Fig. 4. The results of this analysis are summarised in Fig. 5. This figure shows a log-log plot of the singular spectrum of the neighbourhood matrix K' as a function of the radius of the ball defining the

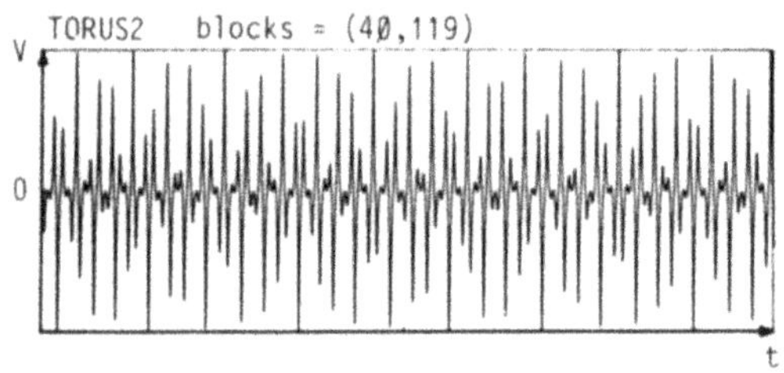

Fig. 1. Time series generated by passing sinusoidal signals through a cubic non-linearity

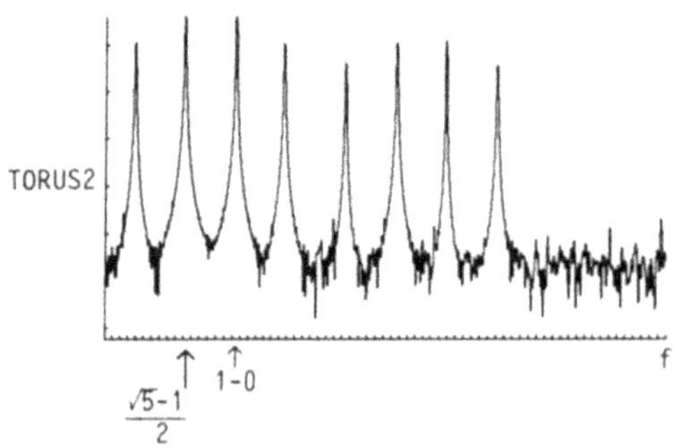

Fig. 2. Power spectrum constructed from the time series shown in Fig. 1

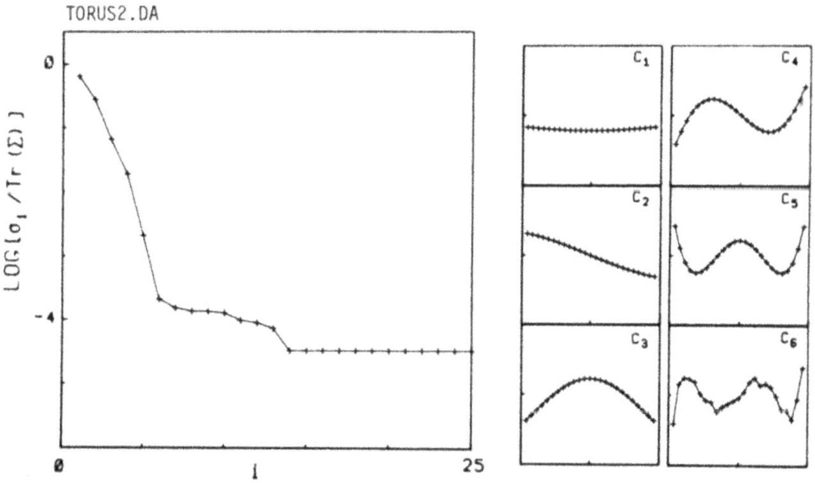

Fig. 3. Global singular value spectrum and associated singular vectors

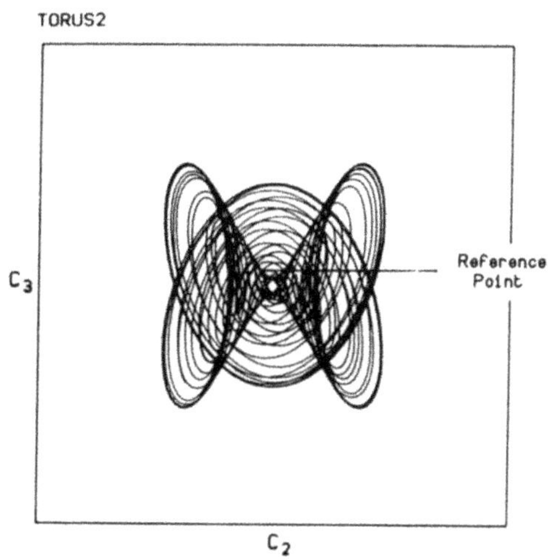

Fig. 4. Phase portrait reconstructed from the time series data shown in Fig. 1. The point marked was used as reference point in the calculation of the topological dimension of the Torus (see Fig. 5)

neighbourhood. We see two singular values which scale linearly with radius, and which are well separated from the rest which do not. This implies that m=2 for this data. This is the expected result. If the neighbourhood can be chosen small enough, then locally the limit cycle would appear planar and would be well approximated by its tangent plane at the point. Thus the singular vector corresponding to the isolated singular value will be in the tangent plane. As the neighbourhood is allowed to increase in size, then providing the local curvature does not vanish we expect a third singular value scaling as r^2 to appear. (The singular vector associated with this will be normal to the curve.) This can be seen in the figure.

Since the singular value that scales as r^2 is apparent at the smallest r we were able to probe with the length of time series used (approximately 10^6 samples), we were never free from the effects of curvature of the manifold. It is to be expected that this situation will worsen as the dimension of the manifold increases.

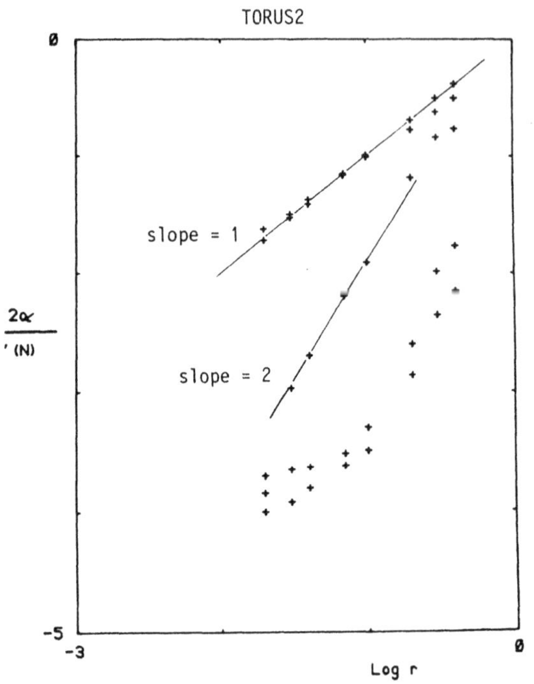

Fig. 5. The local singular value spectrum for the 2-Torus

It is interesting to contrast the results of this local analysis with the Fourier spectra shown in Fig. 2. Due to the nonlinearity the latter show many harmonics which makes it difficult to pick out the number of fundamental frequencies contributing to the spectra. On the other hand the dimension of the manifold determined from the local analysis is exactly two (ie the number of original incommensurate sinusoids).

Another recent interesting result has been discovered in the problem of band-limited imaging by confocal scanning. We consider the one-dimensional case only for simplicity. Here a complex valued function is illuminated by a highly focused laser beam which results in a "windowing" by an Airy function, the Fourier transform of a rectangular spectral band. In the scanning process the window function traverses the object and the resulting windowed values are analysed in a second band-limited system. In any one window position, therefore, we have a set of values g(x) of the original function of (y) given by

$$g(x) = \int_{-\infty}^{\infty} \operatorname{sinc}(x-y)\,\operatorname{sinc}(y)\,f(y)\,dy\ . \tag{22}$$

In [14] it was shown that, if g(x) is detected at only a small number of points, sampled at the Nyquist (Rayleigh) distance, and the method of truncated singular function expansions is used for recovering f(y), then a recovery of the function f(y) can be made with twice the full bandwidth of the window.

f can be represented by the following form

$$f(y) = \sum_{m=-\infty}^{+\infty} f(y_m)\,\operatorname{sinc}[2(y-y_m)] \tag{23}$$

where

$$y_o = 0 \;;\; y_m = \text{sgn}(m)\,(|m| - \tfrac{1}{2}),\; m + \pm 1, \pm 2, \ldots \tag{24}$$

As shown in [14], if we use the sampling expansion for the (noise-free) image g

$$g(x) = \sum_{n=-\infty}^{+\infty} g(n)\,\text{sinc}(x-n) \tag{25}$$

and the expansion (23) for the transmitted part of f, then (22) can be transformed into an infinite dimensional linear system given by

$$g(n) = \sum_{m=-\infty}^{+\infty} A_{nm}\, f(y_m) \tag{26}$$

where

$$A_{no} = \tfrac{1}{2}\delta_{no} \;;\quad n = 0, \pm 1, \ldots \tag{27a}$$

and

$$A_{nm} = \frac{1}{2\pi^2}\,\frac{(-1)^{n+1}}{y_m(n - y_m)} \;;\; m = \pm 1, \pm 2, \ldots \;;\; n = 0, \pm 1, \ldots \tag{27b}$$

Notice that, while the integral operator (21) does not have an inverse but only a generalised inverse, the infinite-dimensional matrix A, given by (27a) and (27b), is invertible since it is equivalent to the restriction of the integral operator (22) to the orthogonal complement of its null space.

We prove in [4] that the inverse matrix A^{-1} has the following simple expression

$$(A^{-1})_{on} = 2(-1)^n \;;\quad n = 0, \pm 1, \ldots \tag{28a}$$

$$(A^{-1})_{mn} = (-1)^{n+1}\,\frac{2n}{n - y_m} \;;\; m = \pm 1, \pm 2, \ldots \;;\; n = 0, \pm 1, \ldots \tag{28b}$$

Once found, the fact that this result satisfies $A^{-1}A = AA^{-1} = I$ can be easily checked by direct computations using value of standard numerical series.

From (13) and (28a) we find that

$$f(0) = 2\sum_{n=-\infty}^{+\infty} (-1)^n\, g(n)\;. \tag{29}$$

This remarkable result obtained in the context of confocal scanning microscopy once again shows the power of the theory of singular systems in the analysis of sequences of sampled data. Here our sequences are in space rather than time, but the principles are the same.

7. Acknowledgements

This research has been partially supported by NATO Grant No 463/84 and EEC Contract No ST29-0089-3.

8. References

1. J.G. McWhirter, E.R. Pike: J. Phys. A 11, 1929 (1978).
2. F. Takens: Lecture Notes in Mathematics, Eds. D.A. Rand and L-S Young (Springer, Berlin, 1981).
3. D.S. Broomhead, R. Jones, G.P. King: J. Phys. A, to be published (1987).
4. M. Bertero, C. De Mol, E.R. Pike: JOSA A, to be published (1987).
5. D.S. Broomhead, R. Jones, G.P. King, E.R. Pike: "Chaos, Noise and Fractals", Eds. E.R. Pike and L.A. Lugiato (Adam Hilger, 1987).
6. C.F. Shannon, W. Weaver: The Mathematical Theory of Communication, (Univ. of Illinois Press, Urbana, 1949).
7. D. Slepian, N. Pollak: Bell System Tech. J. 40, 43 (1961).
8. E. Picard: R. C. Mat. Palermo 29, 615 (1910).
9. E.R. Pike: Inverse Problems, Ed. G. Talenti, Fond CIME Montecatini (Springer-Verlag, Berlin, 1986).
10. H. Whitney: Ann. Maths. 37, 645 (1936).
11. D.S. Broomhead, G.P. King: Physica 20D, 217 (1986).
12. D.S. Broomhead, G.P. King: "Nonlinear Phenomena and Chaos", Ed. S. Sarkar (Adam Hilger, Bristol, 1986), 113.
13. K. Fukunaga, D.R. Olsen: IEEE Trans. Comput. C-20, 176 (1971).
14. M. Bertero, P. Brianzi, E.R. Pike: Inverse Problems, to be published (1987).

Geometrical Principles of Pattern Formation and Pattern Recognition

W. Güttinger, P. Haug, and D. Lang

Institute for Information Sciences, University of Tübingen, D-7400 Tübingen, Fed. Rep. of Germany

1. Introduction

In spite of its complexity our world is not chaos but a ceaseless creation and destruction of forms or structures which are endowed with a degree of stability: they take up some part of space and last for some period of time. One of the central problems of science and, in particular, of synergetics is to explain this change of form and, if possible, to predict it. The formation of spatio-temporal patterns and modes of behavior is first of all a topological problem. Thus, in our quest for understanding how forms are generated in nature,the relationship between physical forces and the stable geometries they can shape is obviously of fundamental importance. This relationship and the ensuing universal character of structure-forming processes are the major themes underlying this paper. Its objective is to set up, in terms of invariant bifurcation theory, a unifying topological framework for spontaneous pattern formation in nonlinear systems and to show that the same geometrical concepts also govern the inverse problem of pattern recognition.

The spontaneous formation of spatio-temporal structures is the result of sudden qualitative changes, known as bifurcations, which occur in a nonlinear system if an externally controllable driving force passes through critical values so that an equilibrium existing between competing system-immanent effects breaks down. As a consequence, an initially quiescent system becomes unstable and, in a sequence of bifurcations, restabilizes successively in ever more complex space- and timedependent configurations. In the vicinity of a bifurcation point a system becomes extremely sensitive to small ambient factors like imperfections, external fields or fluctuations which may lead to symmetry breaking. These enable the system to perceive its environment (e.g., boundary conditions) and to adapt to it by forming preferred patterns or modes of behavior . It is this interplay between multiple bifurcations caused by competing forces, imperfections, symmetry constraints and symmetry breaking effects which produces the complex patterns and changing morphologies that we see in nature. Many of the pattern- forming processes observed in systems of quite different genesis exhibit astonishing analogies which are independent of the composition of the system and of the interactions between individual components. For example, the generation of temporal patterns in neural activity exhibits the same geometrical structure as does the formation of spatial patterns during cellular solidification of an alloy. Such universal characteristics of structure formation point to the existence of general organizational principles. Most prominent among the theoretical programs venturing into the field of pattern formation are Prigogine's dissipative structures [1], Haken's synergetics [2] and Thom's singularity theory [3] which we try to synthesize here, to a first approximation, into a geometrical framework.

Since the formation of structures is a geometrical problem, one expects that the observed qualitative analogies possess a common topological origin. Describing analogies requires an intuitive selection of significant global state variables or order parameters which can be put into correspondence. These quantities measure the degree and type of ordering or structure which is built up or destroyed in the vicinity of a bifurcation point. Spontaneous structural changes are then characterized by the appearance of topological singularities in these observables and a notion of equivalence can be introduced which defines what it means for two bifurcation phenomena to be qualitatively similar, i.e., topologically equivalent. The main result of the ensuing bifurcation theory is that bifurcations are largely determined by geometrical constraints such as the dimension and shape of the system, inherent symmetries and boundary conditions and, furthermore, that the bifurcations producing stable patterns occur in certain definite and classifiable ways. This is a consequence of the fact that physical systems which allow repeatable experiments are structurally stable.

2. Structural Stability

The fundamental laws of physics derive from geometrical invariance principles, such as space-time symmetries, which provide us with conservation laws and equations of motion. Besides, there are material invariance principles, like gauge symmetries, which establish relationships between interactions and the forces that are producing them. The link between geometry and physics is due to a topological invariance principle which relates to the persistence of phenomena. Namely, most physical systems or structures possess an inherent stability property: They preserve their quality under small perturbations, i.e., they are "structurally stable" or robust. Otherwise we could hardly think about or describe them and today's experiment would not reproduce yesterday's result. We do not know how it got that way. But accepting structural stability as a fundamental empirical principle of nature that complements the familiar invariance laws, this principle can be shown to steer the pattern-forming forces and to determine the geometrical rules by which new structures are generated.

The concept of structural stability is invariably tied to a remarkable geometric transversality property: "Two manifolds meet almost always transversely and, therefore, stably." To illustrate this transversality lemma (Thom), consider two intersecting curves C_1 and C_2 in the plane. Intersections can obviously occur in one of two ways: Either (i) transversely, i.e., without tangential contact as in Fig. 1a, or (ii) nontransversely, i.e., with tangential contact as in Fig. 2a. In the case (i) a slight deformation of C_2 by bending into C_2', Fig. 1b, produces qualitatively nothing new: Transversal crossing is insensitive to small perturbation, i.e., structurally stable.
However, in the case (ii) of nontransversal crossing as in Fig. 2a (where C_2 is chosen as a cubic) things are different: Deforming C_2 into C_2' by stretching produces the transverse intersection on top of Fig. 2b. On the other hand, compressing C_2 slightly into C_2'' produces the lower Figure 2b with P bifurcating into three intersections at P'', P''', P''''. Hence, nontransversal crossings are structurally unstable and a small perturbation produces spontaneously a qualitatively new topological situation. The newly generated intersections are, however, stable against further deformations of the

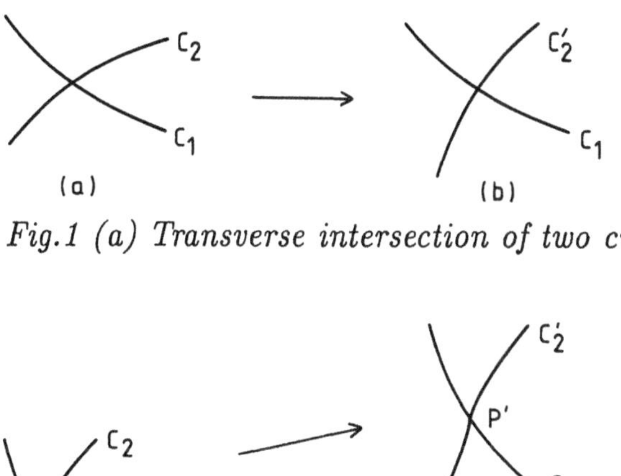

Fig.1 (a) Transverse intersection of two curves, (b) deformation of (a)

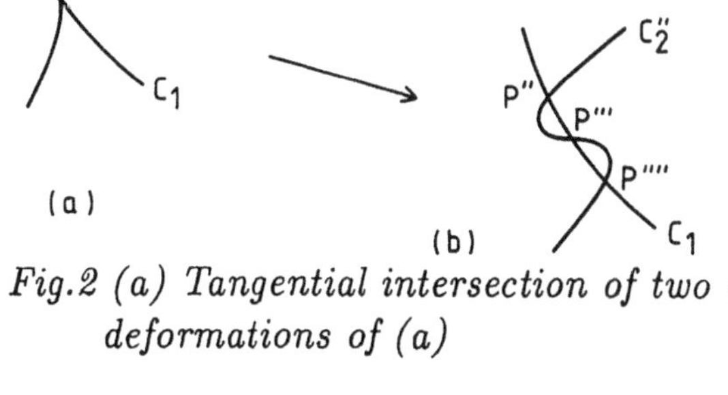

Fig.2 (a) Tangential intersection of two curves, (b) two different deformations of (a)

intersecting curves. The curve C_2 tangent to C_1 at P in Fig. 2 is obviously a singular, in P unstable element embedded into a family of curves C_2', C_2'' which intersect C_1 transversely and stably. Consequently, a bifurcation, i.e., an unstable event, appears as a singular element in a stable process. The transversality lemma generalizes this geometrical situation to hypersurfaces that meet in higher-dimensional spaces: A physical instability is described by a singularity, represented by a nontransversal intersection of hypersurfaces called "organizing center". The degree of contact, called the codimension of the singularity, is a measure of the complexity of the phenomena that spring up when the intersecting surfaces are slightly deformed or "unfolded". By this unfolding or perturbation - generated by varying forces, material or imperfection parameters etc. - the surfaces are deformed in such a way that they intersect transversely and stably. We show below that the intersection curves represent the bifurcation diagrams along which new stable structures can generically evolve. Consequently, bifurcation to stable structures can be attributed to the generation of transverse intersections and pattern formation is reduced to its universal topological root: Bifurcation takes place in a welldefined and definite way when structural stability breaks down in a nontransverse intersection.

To determine the basic bifurcation phenomena for a single order parameter x induced by varying an external control parameter λ and taking account of system parameter α_i, we observe that the family of cubics of Fig. 2 make up the surface shown in Fig. 3, i.e., an overhanging cliff described by $F = x^3 + ux + v = 0$. Setting $u = -\lambda$ and $v = \alpha_1 + \alpha_2\lambda$ defines, for fixed (α_1, α_2), a straight line in the (u, v) - plane as λ varies. Then, cutting the cliff with vertical planes having these lines as basis (Fig. 4) the intersection curves define the bifurcation diagrams

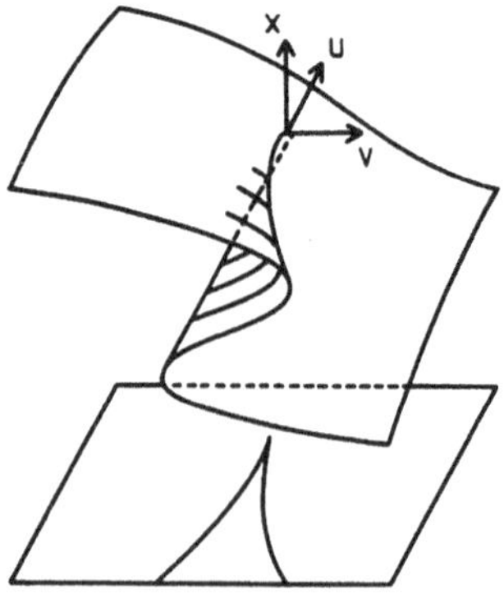

Fig.3 Cusp catastrophe manifold

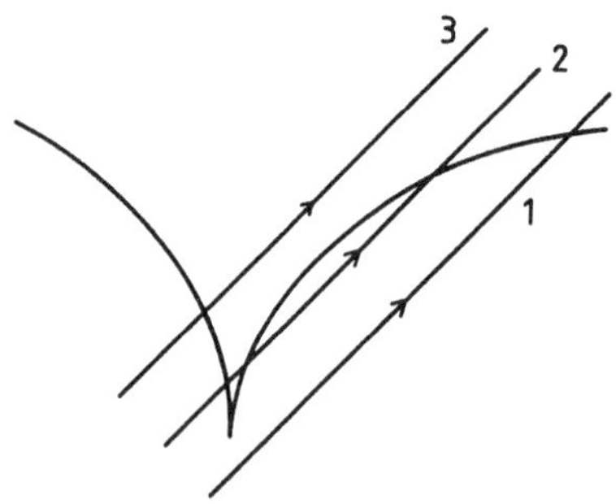

Fig.4 Straight paths through the cusp producing the perturbed pitchfork

of the perturbed pitchfork, i.e., the solution set $x = x(\lambda)$, Fig. 5, of the bifurcation problem

$$F(x, \lambda, \alpha) := x^3 - \lambda x + \alpha_2 \lambda + \alpha_1 = 0 \tag{2.1}$$

for various values of $\alpha = (\alpha_1, \alpha_2)$. Equation (2.1) describes, e.g., the magnetization x of a ferromagnet with temperature $-\lambda$ under the action of perturbations α.

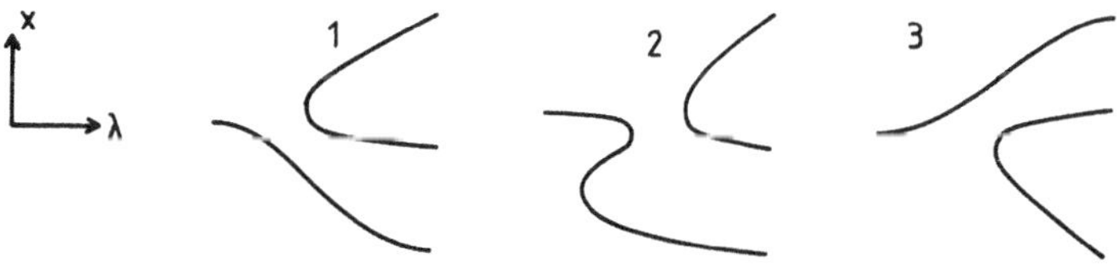

Fig.5 Perturbed pitchfork for various α_1, α_2

Cutting the cliff by vertical surfaces having the parabolae of Fig. 6 as basis yields the hysteresis, mushroom and island bifurcation diagrams of the winged cusp $x^3 + \lambda^2 + \alpha_2 x + \alpha_3 x \lambda + \alpha_1 = 0$ shown in Fig. 7 which are familiar from chemical reaction processes [4], neural activity [5] and problems of optical bistability [6].

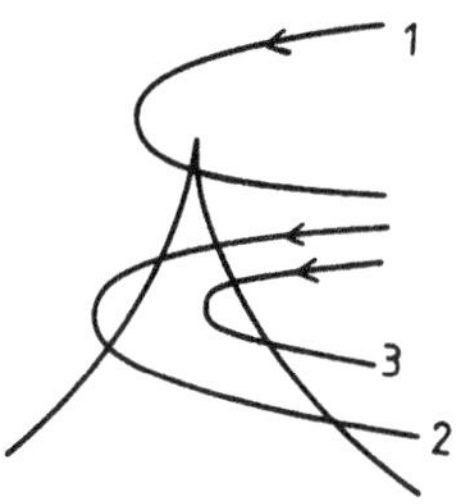

Fig.6 Parabolic paths through the cusp

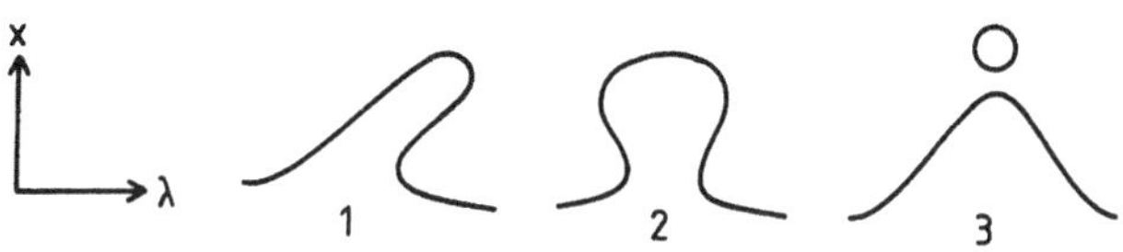

Fig.7 Bifurcation diagrams of the perturbed winged cusp

The above diagrams describe the basic generic bifurcation phenomena obtainable by unfolding the topological singularity represented by the nontransversal cubic intersection of two planar curves. The generalization of these geometrical concepts to the structurally stable unfolding of nontransversely intersecting manifolds with finite codimension generates the topological framework of singularity theory. Traveling through these unfoldings along paths gives rise to invariant bifurcation theory.

3. Bifurcation Theory and Evolution Equations

The objective of this section is to classify transitions between spatio-temporal patterns described by nonlinear evolution equations using unfolded bifurcation equations.

A general bifurcation problem [7] consists of finding the solutions $x = x(\lambda)$ of a system of equations

$$g(x,\lambda) = 0 \qquad (3.1)$$

with $g(0,0) = g_x(0,0) = 0$, where $g = (g_1, ...g_n)$ is a smooth function, $x = (x_1, ...x_n)$ are state variables or order parameters and λ is a distinguished bifurcation parameter representing the control variable in a physical experiment. It may be assumed that (3.1) is the equation system for the amplitudes x of the solution of a given nonlinear evolution equation, obtained, e.g., by a Lyapunov-Schmidt reduction (cf., below). The point $(x,\lambda) = (0,0)$ is a bifurcation point or a singularity of g, and the solutions of (3.1) constitute the bifurcation diagram. Two functions $g(x,\lambda)$ and $g'(x,\lambda)$ are called equivalent or "qualitatively similar" if there exist smooth local coordinate changes $x \rightarrow X(x,\lambda)$ and $\lambda \rightarrow \Lambda(\lambda)$ with det $X_x(0,0) \neq 0$ and $\Lambda_\lambda(0) > 0$ so that

$$g(x,\lambda) = T(x,\lambda) \cdot g'(X(x,\lambda),\Lambda(\lambda)) \qquad (3.2)$$

where T is a square matrix with det $T(0,0) > 0$. Sice Λ may not depend on x, the control parameter λ influences x, but not conversely. Since T is invertible, two equivalent bifurcation problems possess qualitatively the same solution set. The effect of perturbations in a system are incorporated into the bifurcation problem (3.1) by the unfolding of g: A k-parameter function $G(x,\lambda,\alpha)$, with $G(x,\lambda,0) = g(x,\lambda)$ is called an unfolding of g with unfolding or imperfection parameters $\alpha = (\alpha_1, ..., \alpha_k)$. If for any sufficiently small perturbation $\epsilon p(x,\lambda,\epsilon)$ there exists an α so that the perturbed bifurcation problem $g + \epsilon p$ is equivalent (in the sense of (3.2)) to $G(x,\lambda,\alpha)$ and if k is the minimum number of unfolding parameters needed to describe the perturbation, then $G(x,\lambda,\alpha)$ is called a stable or universal unfolding of g, and k is called the codimension of g. The codimension is a measure for the degree of complexity of the bifurcation problem. We have the following theorem: If g has finite codimension then there exists a polynomial $g' = N(x,\lambda)$, called a normal form, which is equivalent to g, and a universal unfolding $G' = F(x,\lambda,\alpha)$ of N which is also a polynomial. Here the new variables X and Λ are denoted again by x and λ. By varying α in $F(x,\lambda,\alpha) = 0$ one obtains a finite number of qualitatively different perturbed or unfolded bifurcation diagrams. The α-parameter space can be divided into a finite number of regions such that for any two parameter values lying within the same region the corresponding bifurcation problems are equivalent and structurally stable. Crossing the boundaries separating these regions produces new qualitatively different diagrams.

Changes in the stability of a solution x of $F = 0$ follow by considering x as equilibrium solution of the system $\dot{x} = F(x, \lambda, \alpha)$. The signs of the real parts of the eigenvalues of the Jacobian of F evaluated at x determine the stability properties of the bifurcating solution branches of $F = 0$. In order that bifurcation occurs in (3.1) and g is equivalent to a normal form, g must satisfy certain degeneracy conditions which express the vanishing of some of its first derivatives and certain nondegeneracy conditions which relate to the nonvanishing of some next higher derivatives. With these techniques it is possible to classify bifurcation problems by their codimension, i.e., to give a list of all possible bifurcation diagrams. The invariant bifurcation theory outlined above can be generalized to problems with symmetry [7]. g is called a bifurcation problem with symmetry group Γ if g is Γ- covariant, i.e., if $g(\gamma x, \lambda) = \gamma g(x, \lambda)$ for all $\gamma \in \Gamma$. g and g' are Γ-equivalent if (3.2) holds with the symmetry conditions $X(\gamma x, \lambda) = \gamma X(x, \lambda)$ and $T(\gamma x, \lambda) = \gamma T(x, \lambda)\gamma^{-1}$. The notions of Γ -covariant unfolding, Γ-codimension etc. are defined in an analogous way. If $g = 0$ is a Γ-equivariant bifurcation problem, a bifurcating solution will have an isotropy subgroup which is not Γ, i.e., the new solution will have less symmetry than the old: The symmetry has broken spontaneously. The following example is instructive. Let $x \in \mathbf{R}$ and suppose that g satisfies the degeneracy conditions $g = g_x = g_{xx} = g_\lambda = 0$, and the nondegeneracy condition $g_{xxx} g_{x\lambda} < 0$ at $(x, \lambda) = (0, 0)$. Then the bifurcation problem $g(x, \lambda) = 0$ is equivalent to the codimension-2 pitchfork $g' = N = x^3 - \lambda x = 0$ with universal unfolding $F = x^3 - \lambda x + \alpha_2 \lambda + \alpha_1$ (Eq. (2.1)). If g possesses gradient form, results of singularity theory can be derived.

The classification of bifurcation processes provided by invariant bifurcation theory allows one to determine and to categorize transitions between spatio-temporal patterns described by systems of nonlinear evolution equations. This is due to a physical principle which leads directly from evolution equations to a bifurcation problem and to the unfolding of an organizing center. Let

$$\frac{\partial u}{\partial t} = L(\lambda) u + N(u, \lambda) \tag{3.3}$$

be a system of evolution equations for u with linear part L and nonlinear part N, u being an element of a Banach space. Suppose, e.g., that, for a critical value $\lambda = \lambda_c$, L possesses eigenmodes Φ_i whose eigenvalues have zero real parts. When λ passes through λ_c the nonlinear term N generates an interaction between the modes $x_i\Phi_i$. As a consequence, the amplitudes x_i of these modes increase indefinitely due to resonance and the system blows up unless the work done by the force N on the i-th mode vanishes, i.e.,

$$g_i(x,\lambda) := < N(u,\lambda), \Phi_i > \equiv \int N(u,\lambda)\Phi_i = 0 \tag{3.4}$$

for all i where $<,>$ denotes the scalar product. Here $u = \sum x_i\Phi_i + W(x,\lambda)$ where $W(x,\lambda)(x = (x_1,...x_n))$ describes the "enslaving" of the stable modes by the unstable ones. This implies that, in order to avoid instability, the system transfers energy from one mode to others thereby changing its course through bifurcation. The bifurcation branches are the curves in (x,λ)-space in which the solution surfaces of (3.4) intersect. When an intersection occurs with tangential contact, then, by varying the imperfection parameters contained in the unfolding of g, the surfaces can be deformed in such a way that they intersect transversely, thereby ensuring structurally stable bifurcation diagrams. Unfolding theory determines the number of these parameters and how they must be inserted into the bifurcation equations on topological grounds, irrespective of how these parameters are related to the material parameters appearing in the evolution equations (3.3). The universality of the bifurcation processes described by (3.1) and (3.4) explains why evolution equations (3.3) of quite different genesis can exhibit the same phenomena. Furthermore, it is obvious that, in general, only very specific relationships between the material parameters contained in N produce the degeneracy conditions g must satisfy for bifurcation to take place. Symmetry constraints on N, e.g., through boundary conditions, carry over to g and reduce the number of bifurcation diagrams. Applications of this formalism include degenerate Hopf bifurcations [8], coupled Hopf and steady-state bifurcations [9] and steady-state mode interactions [10]. Some new applications are discussed in Sec. 5.

The above Lyapunov-Schmidt technique must be supplemented by a center manifold reduction to describe the temporal evolution of the state variables through equations of the form $\dot{x} = g(x,\lambda)$ and by space-dependent Ginzburg-Landau-type equations. Here, emphasis is laid on global bifurcation aspects of vector fields and their unfoldings in the case of bifurcations with multiple degeneracies, taking account of symmetry groups [11]. However, there is so far no general formalism available since no dynamical equivalent of the transversality principle is at hand.

4. Pattern Recognition and Image Processing

The application of bifurcation theory to pattern recognition and image processing systems constitutes an important field of research. The underlying idea is that both pattern formation and pattern recognition should possess common topological roots. A central problem is to design and analyze systems which, by means of sensory and perceptual mechanisms, can construct internal representations of physical structures in the outside world. In order to be action-oriented, these representations should provide descriptions of such structures at different levels of detail. Tomography and biological and computer vision systems are examples.

Consider the inverse problem in remote sensing, viz., the reconstruction of an unknown reflecting surface from backscattered signals [12]. Suppose a point source at x_0 in space emits a spherical wave that is reflected back to x_0 by a smooth two-dimensional surface S. Within the framework of ray geometry we observe that, since source and receiver are at the same place, only the geometrical optics specular reflection points $x_s = x_s(x_0)$ on S, whose distance vectors $\vec{R}_s = x_0 - x_s$ are normal to S, contribute to the echo received at x_0. With $R_s(x_0) = |\vec{R}_s|$ the surface normal is $n = \nabla_{x_0} R_s$ and, therefore,

$$x_s = x_0 - R_s \nabla_{x_0} R_s \tag{4.1}$$

is the equation for the surface. For, as x_0 varies above the surface, each point of S eventually becomes specular. R_s is measured by the two-way traveltime $2\frac{R_s}{c}$, c being the speed of the wave. In an echogram caustic structures are discernible, e.g., those shown in Fig.8 (a seismogram) where R_s is plotted as a function of the horizontal source position. These caustic events and the diffraction patterns around them are the singularities which the layered medium (made up of surfaces S) shown in Fig.9 impresses on the incident wave field. R_s becomes multivalued when x_0 moves along a curve above a concave part of the surface S and crosses the surface's evolute on which neighboring normals touch and focusing occurs.
Consequently, as the source traverses the evolute, the number of rays going through x_0 changes by two and so does the number of arrivals in the echo. The distance $R = |x_0 - x|$ from x_0 to any point x on S obviously has an extremum $R = R_s$ at $x = x_s$, $\vec{t} \cdot \nabla_x R = 0$ for vectors $\vec{t}$ tangent to S. Since the sensing process is structurally stable, singularity theory yields normal forms for R near degenerate specular points. In conjunction with these, (4.1) provides an efficient algorithm for surface reconstruction [13]. The technique can be extended to include diffraction effects which are crucial in edge detection as is evident from the different types of diffraction patterns discernible in Fig.8 [14]. The geometrical method of surface reconstruction

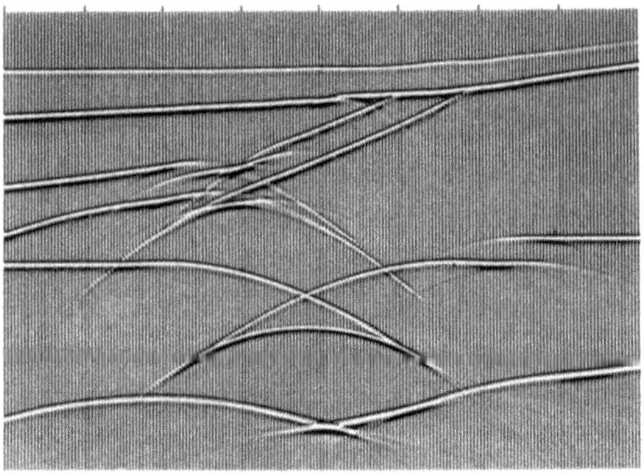

Fig.8 Caustics in an echo recording

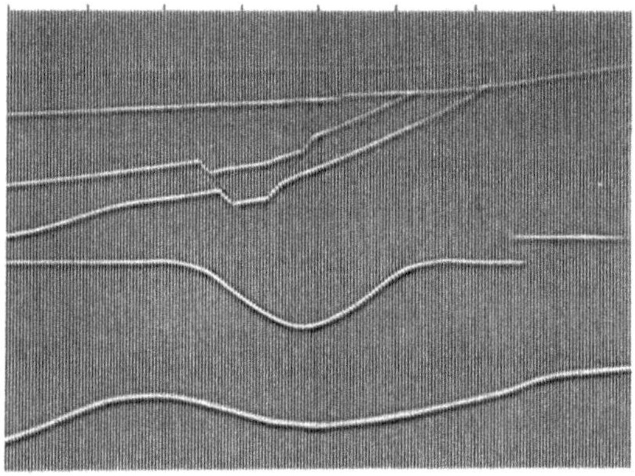

Fig.9 Layered medium producing the caustics in Fig.8

outlined above finds its counterpart in ray-tracing techniques that produce two-dimensional images of three-dimensional objects [15]. Here an object is projected onto an imaging plane by coloring each point in that plane according to the amount of light that reaches the observer from the light sources by passing through the point. To do this, the ray-tracing algorithm traces the paths of light from the observer back to the sources. The caustic bifurcation sets encountered in this computer-graphic process capture the effects of reflection, shadows and hidden surfaces and generate the most realistic images that are currently attainable.

A quite different class of inverse problems consists of reconstructing a surface or function f (smooth or piecewise continuous, or defined on a discrete set) from a large but finite set of incomplete, noisy and possibly ambiguous measured data g: $B(f) = g$ where B is a known operator. In general, the observed g does not determine f in a unique and stable way, i.e., the inversed problem is ill-posed (B^{-1} does not exist) [16]. One redefines the problem by incorporating constraints on f generated by some prior knowledge about its general behavior. Two approaches to these types of problems are presently in the center of interest: (i) regularization by minimizing cost or energy functionals which incorporate a constraint of smoothness of f [17]; and (ii) a probabilistic approach using Bayesian estimation, starting from a Gibbsian probability distribution on the space of possible functions [18]. The resulting algorithms can be implemented by massively parallel hardware, hybrid and neuronal analog networks of Hopfield type [19] that provide fast and next-to-optimal solutions. For example, in the case of a linear inverse problem $Bf = g$, where f and g are vectors and B an $(n \times n)$ matrix with rank $B < n$, f can be determined by minimizing the cost function

$$E(f) = ||Bf - g||^2 + \alpha ||Qf||^2 \tag{4.2}$$

where the second term is a regularizer which renders the problem wellposed. The quadratic form (4.2) is the power dissipated by a linear lumped network in which the components of f represent the voltages. Nonlinear cost functionals that lead to nonlinear networks and fast optimization algorithms can be derived from probabilistic approaches, taking, e.g., the maximum entropy principle as a starting point [20]. Here one has to maximize the entropy $H = -\sum f_i \log f_i$ subject to constraints like

$$C(f) \equiv ||B(f) - g||^2 = \epsilon$$

or, alternatively, find an extremal of $F = H - \alpha C$ for a given α, i.e., the fixed points of

$$f_i = \exp(-\alpha \frac{\partial C}{\partial f_i}).$$

Since the maximum entropy principle is conceptually wellfounded, this supports the suggestion that step-by-step image restoration can be optimally generated in terms of renormalization group operations.

5. Pattern Formation and Mode Interaction

In this section we present new results concerning the interactions of steady state modes near multiple bifurcations in systems with finite spatial extension. Steady state modes branch off a basic state when a distinguished bifurcation parameter passes through a critical value and a number of eigenvalues of the linearized system goes through zero. Variation of an extra parameter may change the eigenvalue structure and lead to multiple bifurcations which induce mode interaction. Paradigms of such interactions are provided by directional solidification of a binary alloy [21] and convection in a fluid [22]. Here, the distinguished parameter λ is the pulling velocity and the Rayleigh number, respectively. The extra parameter is a characteristic length, viz., the finite size of the sample or the aspect ratio. Steady state mode interactions are well understood in systems with one or two spatial dimensions [10], but except for systems with spherical geometries [23] little is known about three-dimensional problems.

Consider a long sample of a dilute binary mixture in the (ξ, η)-plane, which is drawn, at constant velocity v, through a fixed temperature gradient between the hot and cold contacts at A and B (Fig. 10). The temperatures are chosen such that the melt-solid interface of the sample is located between A and B. At low pulling velocity, the interface is planar. As v increases and passes through a critical value v_c, the planar solidification front loses stability and cellular patterns are formed. The system is described by the equations

$$W_{\xi\xi} + W_{\eta\eta} + vW_\eta - W_t = 0 \tag{5.1}$$

$$W = -s - \frac{1}{u} H(s) \tag{5.2}$$

$$s_t + v + W_\eta - s_\xi W_\xi = 0. \tag{5.3}$$

Here W is the solute concentration, $s(\xi, t)$ the position of the interface, u is a material parameter and

$$H(s) = \frac{-s_{\xi\xi}}{(1 + s_\xi^2)^{\frac{3}{2}}} \tag{5.4}$$

the curvature of the interface. Eq. (5.1) holds for $\eta \geq s$ and Eqns. (5.2), (5.3) on the interface $\eta = s(\xi, t)$.

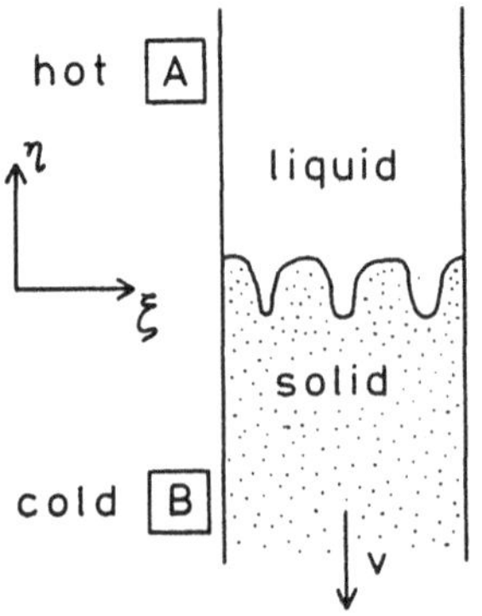

Fig.10 The directional solidification system

Non-flux boundary conditions are imposed at the sidewalls

$$W_\xi = 0 \quad at \quad \xi = 0, b \tag{5.5}$$

where b is the width of the sample, and

$$W \to -1 \quad for \quad \eta \to +\infty \ . \tag{5.6}$$

Setting

$$s = \sum_{m=0}^{\infty} \epsilon_m(t) \cos(m\sqrt{u}k_0\xi) \tag{5.7}$$

where k_0 is the fundamental wave number determined by b, the quasi-stationary approximation to (5.1)-(5.6) yields the infinite set of coupled nonlinear equations for $\epsilon = \{\epsilon_i\}$:

$$\dot{\epsilon}_m = a_m \epsilon_m + \sum_{i=2}^{\infty} M_m^i(\epsilon) \tag{5.8}$$

where M_m^i are homogeneous polynomials of degree i and

$$a_m = (v - 1 - m^2k_0^2)(\frac{v}{2} + (\frac{v^2}{4} + um^2k_0^2)^{\frac{1}{2}}) - v^2. \tag{5.9}$$

This equation gives the neutral-stability curve $\Gamma(u)$ shown in Fig. 11a and the planar interface loses stability when at least one of the a_m becomes positive. Fig. 11b shows that for certain values of u and

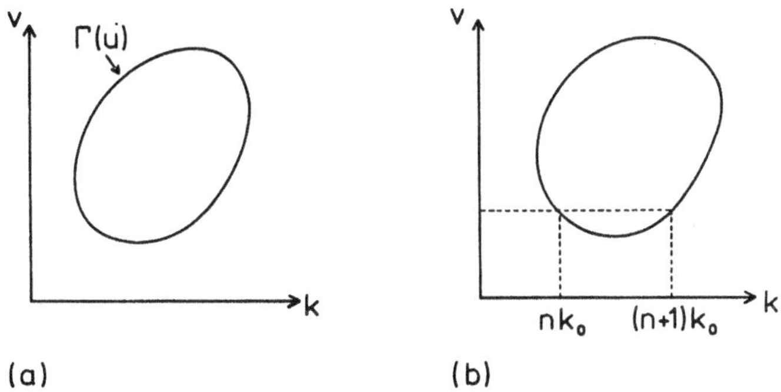

Fig.11 (a) The neutral stability curve Γ, *(b) double-zero eigenvalue*

k_0 two modes with wave numbers nk_0 and $(n+1)k_0$ can simultaneously become unstable as the first ones if v reaches a critical value, i.e., $a_n = a_{n+1} = 0$ and $a_i < 0$ for $i \neq n, n+1$. This gives rise to interactions between the two modes. By a Lyapunov-Schmidt reduction the problem reduces to the solution of two coupled nonlinear equations (3.1) for the amplitudes

$\epsilon_n = x$ and $\epsilon_{n+1} = y$ of the two modes. In virtue of the non-flux boundary conditions (5.5), these equations take the form

$$g_1 = xa(x^2, y^2, v) + x^n y^n b(x^2, y^2, v) = 0$$
$$g_2 = yc(x^2, y^2, v) + x^{n+1} y^{n-1} d(x^2, y^2, v) = 0 \tag{5.10}$$

where a,b, c and d are smooth functions [21]. The form of (5.10) shows that the bifurcation behavior depends on the wave number n.

The solutions of (5.10) can be divided into several classes. The trivial solution $x = y = 0$ describes the planar interface. In the case $n = 1$, a pure-mode solution S_y with amplitude y follows from $x = 0, c = 0$. A mixed-mode solution S_m with amplitudes $x \neq 0$, $y \neq 0$ is given by the solution of $a + yb = 0$, $yc + x^2 d = 0$. If $n \geq 2$, pure-mode solutions S_x with amplitude x follow from $y = 0$, $a = 0$, the pure-mode S_y follows from $x = 0$, $c = 0$ and the mixed-mode is given by $a + x^{n-1} y^n b = 0$, $c + x^{n+1} y^{n-2} d = 0$. S_x and S_y bifurcate from the planar interface while S_m bifurcates in general at secondary bifurcation points and connects the pure-mode branches. The mixed modes give rise to spatial beats and to mode-jumping.

The analysis of (5.10) proceeds as described in Sec.3. The most interesting bifurcation behavior is produced by those values of $u = u_c$ and $k_0 = k_{0c}$ for which $\frac{\partial a(0,0,v)}{\partial v} = 0$ at $v = v_c$ and an analysis shows that the unfolding parameters α, β can be identified with $u - u_c$ and $k_0 - k_{0c}$ [21]. A codimension two bifurcation point, corresponding to the degeneracy condition $a_v(0, 0, v_c) = 0$, exists for all $n \geq 1$. For $n = 1$ a typical normal form is given by

$$-x(y + \lambda^2 + \alpha + \beta\lambda) = 0$$
$$x^2 + y(y^2 + \lambda) = 0 \tag{5.11}$$

where $\lambda = v - v_c$. A corresponding bifurcation diagram is shown in Fig.12 in which the signs denote those of the real parts of the eigenvalues of the Jacobian of (5.11). Here a tertiary bifurcation point (TB) is discovered which describes a Hopf bifurcation to a temporally oscillating interface as a result of the interaction between two stationary modes. Such oscillations have recently been found in numerical simulations [24]. A typical bifurcation diagram for $n = 2$ is shown in Fig.13.

The equations of motion (5.1) - (5.4) are readily generalized to describe 3-dimensional solidification of a sample of binary alloy with rectangular cross section. In this case we have now two fundamental wave numbers k_1 and k_2 corresponding to the ξ- and ζ- axes of the sample and the curve $\Gamma(u)$ of Fig.10 expands into a torus $T(u)$ in (k_1, k_2, v)-space which is invariant under rotation of $(K_1, K_2) = (mk_1, nk_2)$. When v exceeds a critical value

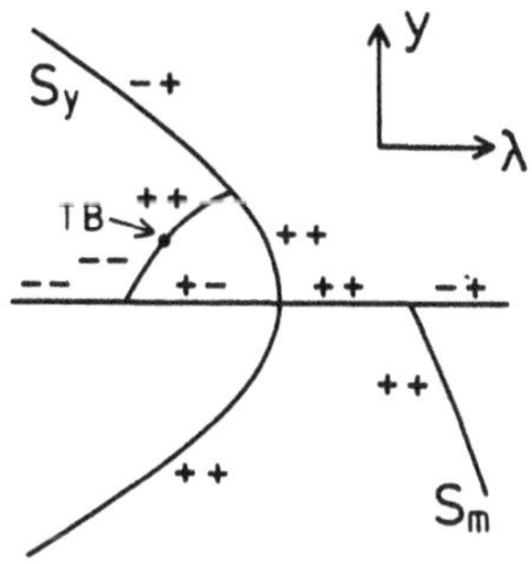

Fig.12
Diagram of normal form (5.11) for special values of α and β. TB denotes a tertiary Hopf bifurcation.

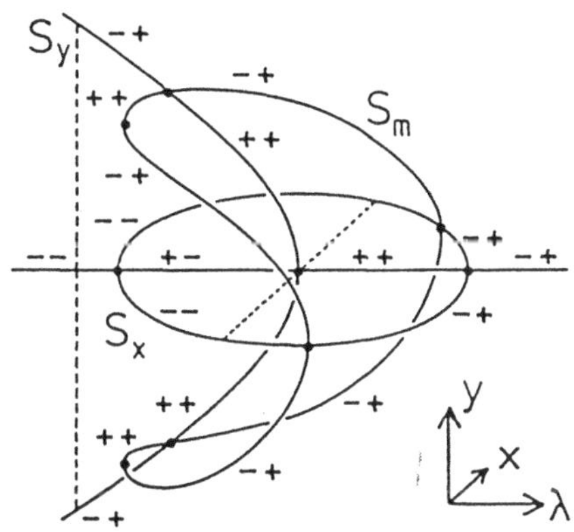

Fig.13 Diagram for n=2

v_c, the surface $T(u)$ intersects the plane v =const in (K_1, K_2, v)-space in two circles with radii $R_i(v) < R_o(v)$ such that

$$a_{mn} = (v - 1 - K_1^2 - K_2^2)(\frac{v}{2} + (\frac{v^2}{4} + u(K_1^2 + K_2^2))^{\frac{1}{2}}) - v^2 \qquad (5.12)$$

(which generalizes (5.9)) is positive in the sector between the two circles and negative outside. Now, the non-flux boundary conditions at the sidewalls force K_1 and K_2 to lie on a rectangular grid spanned by the vectors $(k_1, 0)$ and $(0, k_2)$. Hence, bifurcation occurs if one ore more such lattice points are located on the circles and all others stay outside the sector between R_o and R_i. The condition that a pair (K_1, K_2) is on a circle leads to $K_1^2 + K_2^2 = R^2$ where $R = R_o$ or R_i. For rationally dependent K_1, K_2, R this condition reduces to the determination of Pythagorean triples. From this it follows that, by suitable choices of k_1 and k_2, any prescribed number of modes can become simultaneously unstable as v passes through v_c. The bifurcation analysis proceeds as above with

$$\dot{\epsilon}_{mn} = a_{mn}\epsilon_{mn} + \sum M^i_{mn}(\epsilon)$$

taking over the role of (5.8), ϵ_{mn} representing the amplitudes in a double-cosine series describing the interface $s(\xi, \zeta, t)$.

Since, as we have seen, 3-dimensional solidification is governed by multi-mode instabilities, a classification scheme for the normal forms describing the bifurcation phenomena must make use of group theoretical arguments . To illustrate the results let us assume that three modes become unstable simultaneously. This yields three coupled nonlinear equations for the amplitudes $(x, y, z) = (\epsilon_{m_1 n_1}, \epsilon_{m_2 n_2}, \epsilon_{m_3 n_3})$. It can be shown that, if the three mode instability has codimension three (i.e., if three parameters are fixed), then these equations take the normal forms

$$\begin{aligned} x(x^2 + ky^2 + pz^2 - \lambda - \alpha) &= 0 \\ y(ex^2 + y^2 + gz^2 - \lambda - \beta) &= 0 \\ z(fx^2 + ey^2 + z^2 - r\lambda) &= 0 \end{aligned} \tag{5.13}$$

where $\lambda = v - v_c$, α and β are unfolding parameters and the remaining coefficients are the modal parameters which depend on u, k_1 and k_2. As explained in Sec.3, the (α, β)-plane can be divided into a number of regions such that in each region a structurally stable bifurcation diagram prevails. As an example we show in Fig.14 for $u = 27.5$ a schematic bifurcation diagram for the modes $(m, n) = (7, 0), (0, 5)$ and (4,4) with single-mode branches S_x, S_y, S_z (which are defined by $y = z = 0, x = z = 0, x = y = 0$, respectively), and two-mode solutions S_{xy} etc. determined by one zero and two nonzero amplitudes. The diagram shows both hysteresis and mode-jumping. The form of the interface (a contour plot) for an S_{xz}-solution is shown in Fig.15.

We have chosen directional solidification as a paradigm for steady state mode interaction, but the formalism also applies to other physical systems, e.g., to Rayleigh-Bénard convection. To conclude with, we discuss briefly a result of a simulation experiment of the Boussinesq approximation describing convection in a rectangular box. The system is described by equations

$$\begin{aligned} \frac{1}{\sigma}(\nabla^2 \psi_t + J(\psi, \nabla^2 \psi)) &= R\Theta_x + \nabla^4 \psi \\ \Theta_t + J(\psi, \Theta) &= \psi_x + \nabla^2 \Theta \end{aligned} \tag{5.14}$$

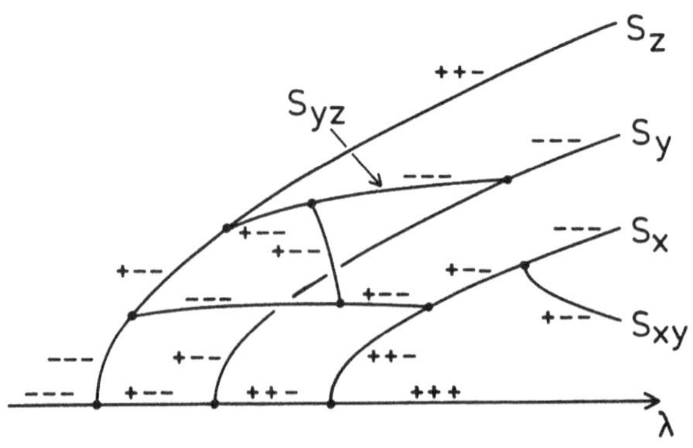

Fig.14
Bifurcation diagram of (5.13) for $u = 27.5$ and special values of α and β

Fig.15
Contour plot of a S_{xz}-interface

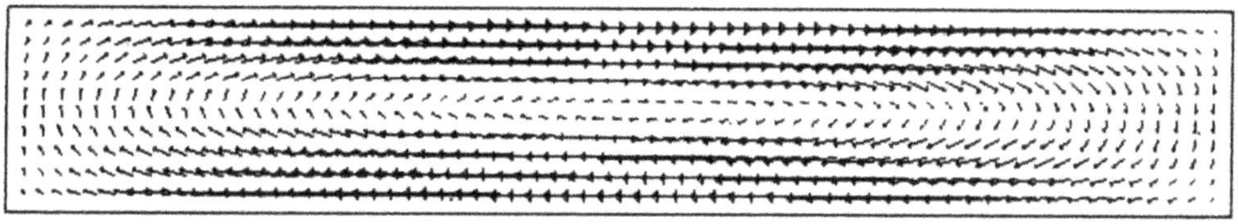

t = 0

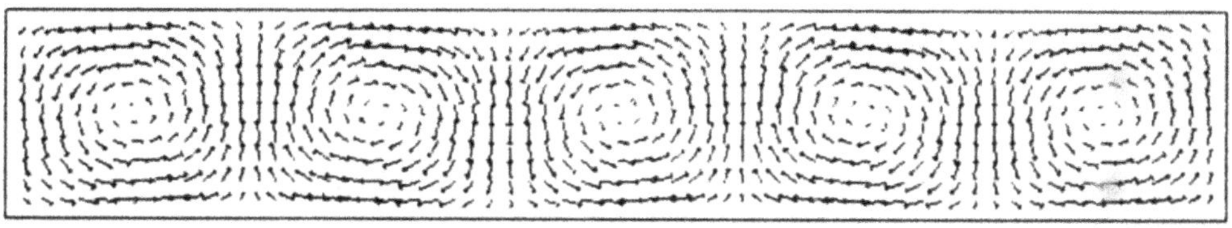

t = 3500

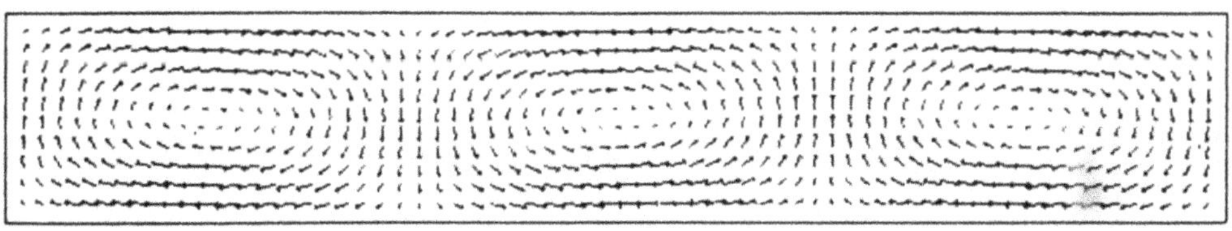

t = 10000

Fig.16 Time-evolution towards a stable three-roll configuration starting from a 1-roll initial state ($R = 820$, $l = 6.0$, $\sigma = 0.1$)

where $J(f,g) = f_x g_z - f_z g_x$ and the boundary conditions

$$\begin{aligned} \psi = \psi_{zz} = \Theta = 0 \quad & for \quad z = 0,1 \\ \psi = \psi_{xx} = \Theta_x = 0 \quad & for \quad x = 0,l \, . \end{aligned} \tag{5.15}$$

Here ψ is the stream function, Θ the deviation of the temperature from the basic conducting state, R is the Rayleigh number and σ the Prandtl number. Eq. (5.15) are the simplest boundary conditions, with the temperature fixed at the top and the bottom of the box, and no sideways heat flux and no tangential viscous stresses [22]. The basic bifurcations are here transitions between convective roll patterns. To simulate Eqns. (5.14),(5.15) we discretize these on a rectangular grid and represent them by cellular automata. Starting from an initial configuration and varying the parameters, the system evolves in time and ends up in some stable configuration. As an example, Fig.16 shows how an unstable one-roll pattern evolves into a stable three-roll pattern through an intermediate five-roll configuration.

6. Conclusion

We have shown, in terms of representative examples, that the topological singularities and bifurcation processes deriving from the principle of structural stability determine the dominating phenomena and features observed in both structure formation and pattern recognition. Because of their universality and classifiability, these geometrical bifurcation processes also provide a basis for our understanding of the qualitative analogies that have been discovered in the critical behavior of nonlinear systems of various genesis whose unifying description is one of the central aims of synergetics. Beyond the bifurcation theory of systems with few order parameters discussed above lies the terra incognita of nonlinear large-scale systems of very high dimensionality, such as neural networks and probabilistic computers with selforganisational properties, whose dynamics is not yet amenable to an analytic description. The robustness of such systems and the observed sudden changes between collective patterns, however, lead one to ask for organization principles that generalizes those discussed in this paper. These problems raise difficult but fascinating questions that promise a great challenge for future research.

Acknowledgments

It is a pleasure to acknowledge helpful discussions with G. Dangelmayr, Ch. Geiger and M. Neveling, and the support by the Stiftung Volkswagenwerk.

References

[1] G. Nicolis and I.Prigogine: *"Selforganization in Nonequlibrium Systems"*, Wiley (1977)

[2] H. Haken, *"Synergetics"* and *"Advanced Synergetics"*, Springer (1983)

[3] R. Thom, *"Structural Stability and Morphogenesis"*, Benjamin (1975)

[4] M. Golubitsky and W. Langford, *J. Diff. Equs. 41* , 375 (1981)

[5] I. Labouriau, in Ref. [7]

[6] D. Armbruster, *Z. Phys. B53*, 157 (1983)

[7] M. Golubitsky and D. Schaeffer, *"Singularities and Groups in Bifurcation Theory"*, Springer (1985)

[8] see Ref. [7]

[9] D. Armbruster, G. Dangelmayr and W. Güttinger, *Physica 16D*, 99 (1985)

[10] G. Dangelmayr and D.Armbruster, *Contemp. Math. 56*, 53 (1986)

[11] J Guckenheimer and P. Holmes, *"Nonlinear Oscillations, Dynamical Systems and Bifurcations of Vector Fields"*, Springer (1985)

[12] G. Dangelmayr and W. Güttinger, *Geophys. J. R. Astro. Soc. 71*, 79 (1982)

[13] M. Brown and F. Tappert, *Geophys. J. R. Astro. Soc. 88*, 217 (1987)

[14] W. Güttinger in *"Frontiers in Nonequilibrium Statistical Physics"* ed. by G. Moore and M. Scully, Plenum (1986)

[15] M. Henderson in *"Bifurcation: Analysis, Algorithms, Applications"* ed. by T.Küpper, R. Seydel and H. Troger, Birkhäuser (1987)

[16] W. Boerner et al. (ed.), *"Inverse Methods in Electromagnetic Imaging"*, Reidel (1985)

[17] T. Poggio and C. Koch, *Proc. R. Soc. Lond. B226*, 303 (1985)

[18] S. Geman and D.Geman, *IEEE Trans. PAMI 6*, 721 (1984); C. Koch, J.Marroquin and A. Yuille, *Proc. Nat. Acad. Sci. USA*, 1986 (in press)

[19] J. Hopfield, *Proc. Nat. Acad. Sci. USA 81*, 3088 (1984); J. Hopfield and D. Tank, *Biol. Cybern. 52*, 141 (1985)

[20] E. Jaynes, *Proc. IEEE 70*, 939 (1982)

[21] P. Haug, *Phys. Rev. A35*, 4364 (1987); C. Geiger, W. Güttinger and P. Haug in *"Complex Systems - Operational Approaches"*, ed. by H. Haken, Springer 1985

[22] E. Knobloch and J. Guckenheimer, *Phys. Rev. A27*, 408 (1983)

[23] R. Friedrich and H. Haken in *"The Physics of Structure Formation"* ed. by W. Güttinger and G. Dangelmayr, Springer, 1987

[24] M. Bennett, R. A. Brown and L. Ungar in *"The Physics of Structure Formation"* ed. by W. Güttinger and G. Dangelmayr, Springer, 1987

Part IV

Spin Glass Models and Neural Networks

Dynamics of Spin Glasses and Related Models of Neural Networks

H. Horner

Institut für Theoretische Physik, Universität Heidelberg,
Philosophenweg 19, D-6900 Heidelberg, Fed. Rep. of Germany

Abstract: Retrieval of information from a neural network, used as auto-associative memory, is viewed as a dynamic process built from a linear mapping and a local nonlinear rectification. Two types of dynamics are compared: Glauber or Monte Carlo dynamics, the process usually adopted for spin glasses, and dynamics based on differential equations for the generating potentials of neurons. For the second type retrieval takes place via successive refinement and this yields an improved performance in the reconstruction of patterns.

1. Introduction

The brain has been discovered by a growing community of physicists as an object of their investigations. This is partially driven by the hope that the brain, being an assembly of a huge number of neurons, can be approached by methods of statistical physics, especially those developed for disordered systems, e.g. spin glasses, for nonlinear dynamics, phase transitions and self-organizing structures. It is clear that such an approach has to be based on very crude and simplified modelling of the actual processes in assemblies of neurons including their synaptic connections. Furthermore, only some elementary functions might be captured. Among those the functioning as an auto-associative memory is studied best and will be the topic of this and part of the following talks.

The purpose of such a memory is to learn certain patterns, coded as configurations of active (firing) and inactive (refractory) neurons, and to retrieve them if part of them is given as a key. For instance, the pattern "rose" may comprise typical images of roses, their smell and the word "rose". The smell as a key should then be sufficient to retrieve the whole pattern.

Such a model, which triggered most of the present activities among physicists, was proposed by Hopfield [1]. The phase space of this model contains the 2^N states represented by N values $\sigma_i = \pm 1$ with $i = 1...N$. The "Ising spins" σ_i describe McCulloch–Pitts neurons [2], $\sigma_i = 1$ stands for an active, $\sigma_i = -1$ for an inactive neuron. Learning is done by appropriate choice of the efficiencies J_{ij} of the synapses of neuron j terminating at neuron i. J_{ij} can take positive and negative values. If $J_{ij} = J_{ji}$, an energy

$$E = -\frac{1}{2}\sum_{ij}\sigma_i J_{ij}\sigma_j \tag{1.1}$$

can be attributed to each state. Furthermore, a dynamical process has to be prescribed ruling the motion in phase space. It is assumed to be asynchronous sequential in Hopfield's model and parallel in the model earlier proposed by Little [3]. The J_{ij} are chosen such that the energies are local minima for the learned patterns. Local is defined with respect to the dynamics. If the dynamics is such that the energy is lowered in each step, at least in the average, the patterns are attractors of the system.

Elements of this model go back much further. The representation of the activity by $\sigma_i = \pm 1$ has been mentioned. The appropriate choice of the synaptic efficiencies J_{ij} (adaption) originates in principle from Pavlov and his dog, was developed further by Hebb [4], Steinbruch [5], Cooper [6] and others, and analyzed in great detail by Kohonen [7].

Regarding the dynamics I would like to mention a different approach contained for instance in Grossberg's work [8] based on equations of motions for the generating potentials. It models the physiological facts of neurons much better than the Glauber dynamics [9] used in Hopfield's model. The latter is appropriate for magnets for instance, but appears to have no direct support from the physiology of neurons. Later in this talk I will show that the performance of such an auto-associative memory can be quite different for both dynamics. Under certain circumstances the generating potential dynamics appears superior.

The Hopfield model has been analyzed, modified and improved in many different directions [10]. Methods originally invented for spin glasses have provided estimates of the storage capacity [11, 12] of the original model and of variants of it. Correlated patterns can be handled [12, 13, 14] and sequences of patterns can be stored [15], both based on generalizations of Kohonen's work [7]. The robustness of the model has been tested with respect to dilution of synapses [16, 17] or learning by simply clipping off synapses [16, 18 – 20]. Learning and forgetting [21, 22] has been modelled as well as hierarchical storage and retrieval [23]. This is only an incomplete list of recent advances and more will be given in the following contributions.

This development is not only of interest to our understanding of the functioning of neural nets, but has also its impact on the design of parallel computers [24]. Several machines of this type have been built.

In this contribution I will only be concerned with the relatively simple question of how to store a small number of patterns, try to explain the principles and present some new thoughts on the performance of such a network comparing the retrieval with the two dynamics mentioned earlier [25].

I am aware of the fact that an enormous wealth of knowledge has been accumulated by researchers not mentioned so far, but I am still learning and I am not in a position to give proper credits to those.

2. Spin Glasses

Concepts and methods developed for spin glasses [26] have greatly affected the development mentioned above. Spin glasses are disordered magnetic materials. A relatively simple model, the SK–model [27], is described by an energy (1.1). The "exchange constants" J_{ij} are assumed to be random Gauss–distributed variables with zero mean. Single spin flip or Glauber dynamics [9] is employed. Figure 1 shows the energy of such a model with only 8 spins. The abscissa is a one–dimensional projection of the phase space and each node

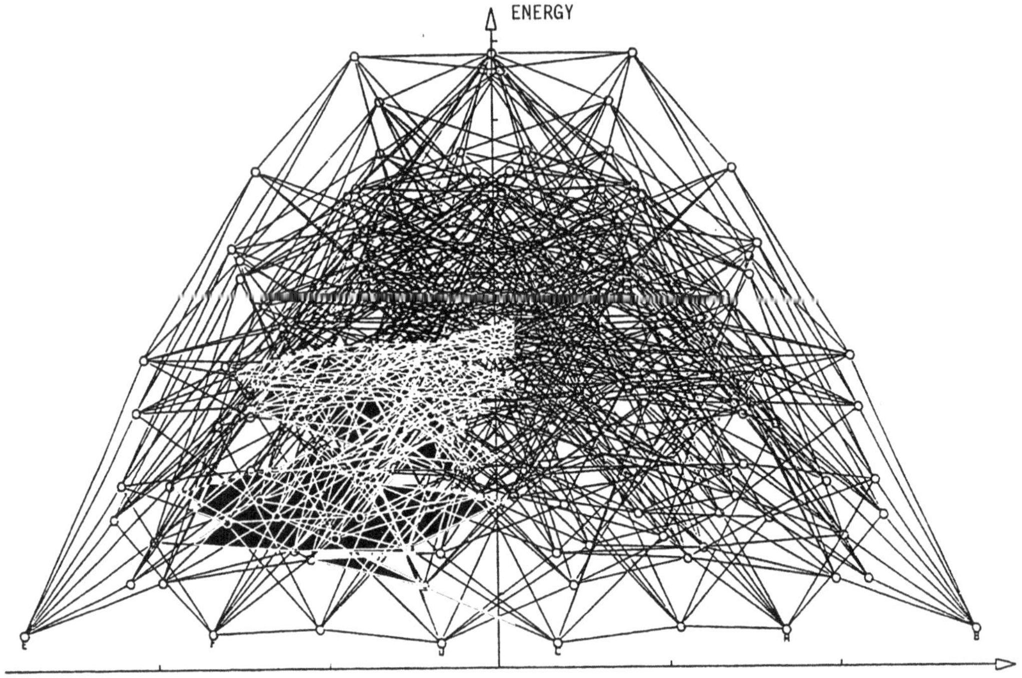

Fig. 1: Energies of an SK-spinglass with 8 spins.

represents one of the 2^N states. Neighbouring states in terms of the dynamics, i.e. states which agree in all except one spin, are connected.

The system performs at finite temperature a random walk along those lines preferring low-energy states at low temperature. In the thermodynamic limit $N \to \infty$ and at temperatures below some critical value at least some of the times required to go from one minimum to another diverge, the system becomes nonergodic. There has been considerable effort over the last years in order to understand this low-temperature phase [26]. Progress has certainly been made, but to which degree this phase has been understood is controversial [28]. This is, however, a different story.

The extraordinary interest in spin glasses originates from the fact that they represent a prototype for a large variety of systems with competing and quasi-random interactions.Models of neural nets are in the same class. In some sense they appear, however, to be more well behaved. Other examples are certain optimization problems of technical relevance, for instance in chip design, relaxations in the ternary structure of proteins and models of evolution, in order to name only a few of them. The idea behind the Hopfield model is to choose the J_{ij} such that the learned patterns become local minima of the energy and to choose the dynamics such that any initial state relaxes toward one of the minima.

Even if the number of stored patterns is small, other local minima appear, so-called mixture states. Their energy is higher than that of the patterns, but their number is much larger than the number of stored patterns [11]. They represent a problem in the Hopfield model and in generalizations of it.

3. Linear Maps and Local Rectification

Essentially all models which have been proposed as associative memories contain two basic steps, a linear mapping and a local rectification process. The distinction of these two steps allows a quite general discussion.

3.1 Linear maps

Consider a neural net with N neurons. The couplings J_{ij} denote the efficiencies of the synapses of neuron j attached to neuron i. They are the components of an $N * N$ matrix J. The state of the net, assuming McCulloch–Pitts neurons, is represented by an N component column vector σ with elements $\sigma_i = \pm 1$. The mapping

$$h = J\sigma \qquad \text{or} \qquad h_i = \sum_j J_{ij}\sigma_j \tag{3.1}$$

defines an "output" vector with components h_i for each "input" σ.

Assume the network is supposed to store $p < N$ linearly independent patterns X^μ with components $\xi_i^\mu = \pm 1$. They may be arranged as an $N * p$ matrix X with columns X^μ. If we can find J such that

$$JX = X, \tag{3.2}$$

the patterns reproduce themselves. This is obviously a minimal requirement for any auto-associative memory. Equation (3.2) is solved with a $p * N$ matrix $\hat{X}$, having elements $\hat{\xi}_\mu^i$, such that

$$\hat{X}X = I_p \qquad J = X\hat{X} \tag{3.3}$$

where I_p is the p-dimensional identity matrix.

There are actually many solutions because eq. (3.3) represents p^2 linear equations for the pN unknowns $\hat{\xi}_\mu^i$. Kohonen [7] has derived criteria for the optimal solution if a linear map, without the rectification discussed later, is used as an auto-associative memory. In this case J has to be chosen as a symmetric projection operator onto the subspace spanned by the patterns and $\hat{X}$ is the pseudo-inverse of X. For this choice the output h of any input state σ is a linear combination of the patterns.

The p-component vector

$$m = \hat{X}\sigma \tag{3.4}$$

or its components m_μ are measures of the similarity of an input state σ to a state μ. These quantities are called overlaps and for a pattern X^μ eq. (3.4) yields $m_\mu = 1$ and $m_\nu = 0$ for $\nu \neq \mu$. For the special case of orthogonal patterns

$$\frac{1}{N}\sum_i \xi_i^\mu \xi_i^\nu = \delta_{\mu\nu} \tag{3.5}$$

we find $\hat{\xi}^i_\mu = \xi^\mu_i/N$ or

$$J_{ij} = \frac{1}{N}\sum_\mu \xi^\mu_i \xi^\mu_j \tag{3.6}$$

which is the form used for instance in the Hopfield model.

3.2 Rectification

An ideal auto-associative memory should, for a given input state, select the particular pattern which has the largest overlap. If the linear map (3.1) is used as a first step, the output h in general does not have the "format" of a state, i.e. its components are not restricted to ± 1, unless the input is a complete pattern. A second step is necessary. This rectification serves also as an error correction device and is ultimately responsible for the robustness of the net. An obvious choice for this rectification is

$$\sigma_i = \text{sign } h_i \tag{3.7}$$

or if the requirement $\sigma_i = \pm 1$ is somewhat relaxed,

$$\sigma_i = \text{th } \beta h_i. \tag{3.8}$$

In the corresponding spin-glass model β would be the inverse temperature and h_i the local field.

Note that the rectification is a local process. This means the summation in the evaluation of the local fields (3.1) and the rectification (3.7,8) can be done on each neuron independently and in parallel.

Auto-associative memories where the linear map and the rectification are performed only once, work well and efficiently, as demonstrated by Palm [29], if the typical patterns have low activity, i.e. if only a small fraction of neurons is active. This actually appears to be the case in our central nervous system, but the activity is higher than required for an optimal functioning of Palm's scheme.

In Palm's work we can find an interesting sideremark. The simplest task realizable on a computer would be to compute the overlaps directly from (3.4) and to select the largest one. This would be the most efficient procedure with respect to the number of elementary operations required and it could be done more or less in parallel. Only if the activity is low enough, a network of the type described here is more efficient.

For higher levels of activity the performance of the net is improved if the basic steps, linear map and rectification, are performed iteratively. There are, however, many possibilities. In Little's model [3] the rectification is done at all neurons simultaneously leading to parallel synchronous dynamics. In Hopfield's model rectification is performed only on a single neuron, selected at random, between two successive mappings. This is a sequential stochastic type of dynamics. The performance of both dynamics appears to be similar.

There is yet another obvious possibility which is, to my knowledge, not discussed in the literature. This is to select the neuron with the largest local field pointing in opposite direction, in other words to select the maximum of $-h_i\sigma_i$ and to update the corresponding neuron. This type of dynamics will be discussed later in more detail and will be motivated from equations of motion for the generating potentials of the neurons.

3.3 Learning

A complete model has to contain learning rules. This means initial states for the couplings J_{ij} have to be specified [30] and a prescription has to be given how the J_{ij} are changed if some pattern is presented. This rule should be local according to Hebb [4] for neural nets. For technical applications [13] this constraint is obsolete. The Hopfield prescription (3.6) can easily be implemented locally [1], but even for Kohonen's prescription [7, 13] iterative local implementations have been proposed [14].

3.4 An example

It is perhaps useful to demonstrate an example, simple enough to be done without a home computer on a piece of squared paper. Assume a net with 64 neurons arranged in a square. The three orthogonal patterns shown in Fig. 2 are stored. Black means an active, white an inactive neuron.

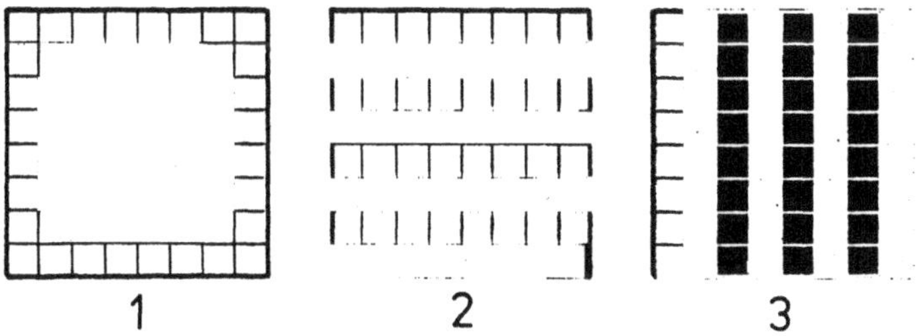

Fig. 2: Three orthogonal patterns with $N = 64$.

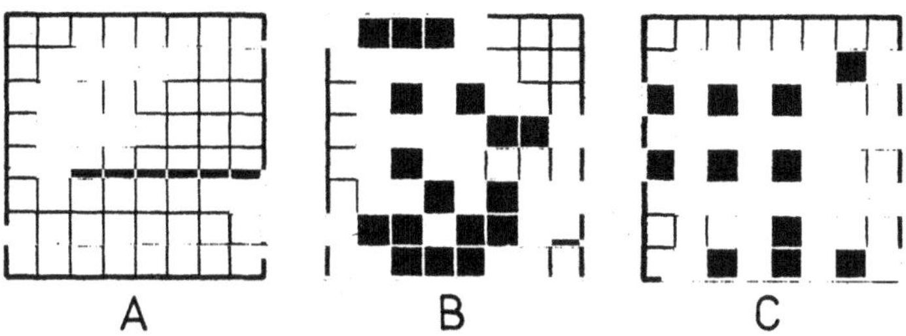

Fig. 3: Input states (A, B) and mixture state (C).

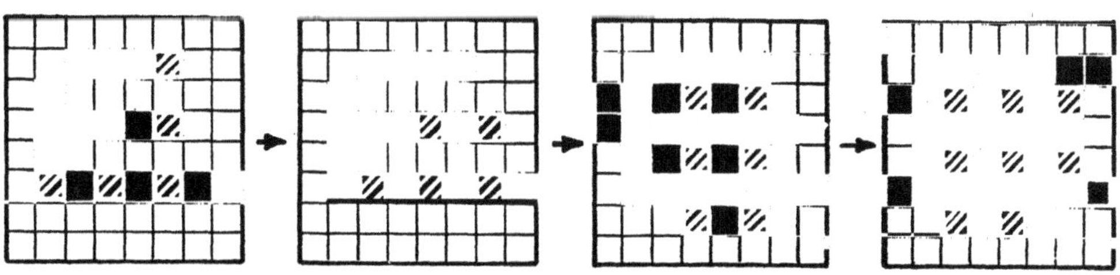

Fig. 4: Evolution of state A toward pattern 1 under the maximum field dynamics.

The state A shown in Fig. 3 is the input. The intitial overlaps are $m_1 = 14/64$, $m_2 = 10/64$ and $m_3 = 6/64$. According to the Little or Hopfield dynamics it evolves toward the state C of this figure. This is obviously a mixture state and contains elements of all three patterns.

Under the "maximum field" dynamics it evolves toward the first pattern. The reconstruction takes place in four steps shown in Fig. 4.

1) Neurons which are on in all three patterns are turned on. This yields $m_1 = 24/64$, $m_2 = 20/64$ and $m_3 = 16/64$.

2) Additional neurons which are on in patterns 1 and 2 but off in 3 are turned on with the result $m_1 = 34/64$, $m_2 = 30/64$ and $m_3 = 2/64$.

3) Neurons which are on in 1 and 3 and off in 2 are turned on and $m_1 = 46/64$, $m_2 = m_3 = 18/64$.

4) Pattern 1 is finally recognized and the "mistake" in the input state is corrected.

Contrary to Little's dynamics where the mixture state is reached in a single step, the maximum field dynamics yields a stepwise recognition where the most common features are detected first and less probable assumptions are successively thrown away, however, not without "rethinking" (step 3).

A second input B is shown in Fig. 3. It has the same initial overlap as the state A and it evolves for all three types of dynamics toward the mixture C. Apparently the system is confused even if the maximum field dynamics is used. We come back to its interpretation in section 5.

4. Glauber–Mean Field and Generating Potential Dynamics

A direct evaluation of the stochastic Hopfield dynamics or the corresponding finite temperature Monte Carlo dynamics [11,26] is difficult and involves the discussion of a master or Fokker–Planckequation for probability distributions. If the number of patterns is low enough [11], it is sufficient to investigate the dynamics of averages.

4.1 Glauber–mean field equations

The resulting dynamical equations are

$$\dot{\sigma}_i(t) = -\sigma_i(t) + \text{ th } \beta\, h_i(t) \tag{4.1}$$

where the local field is given by (3.1). The variables $\sigma_i(t)$ are now interpreted as averages and can change continuously in time. The above equations can also be interpreted as genuine dynamics for neurons with graded response. For a constant local field $\sigma_i(t)$ relaxes toward the value prescribed by (3.8).

It is easy to derive the resulting equations for the overlaps m_μ, eq. (3.4). They can be cast into the form

$$\sum_{\nu} c^{\mu\nu} \dot{m}_{\nu} = -\delta F(m)/\delta m_{\mu} \tag{4.2}$$

where

$$c^{\mu\nu} = \sum_{i} \xi_i^{\mu} \xi_i^{\nu} \tag{4.3}$$

is the correlation matrix and

$$F(m) = \frac{1}{2} \sum_{\mu\nu} m_{\mu} c^{\mu\nu} m_{\nu} - \frac{1}{\beta} \sum_{i} \ln \text{ ch } \beta \sum_{\mu} \xi_i^{\mu} m_{\mu}. \tag{4.4}$$

This function, introduced by Amit *et al.* [11] for orthogonal patterns and generalized by Kanter and Sompolinsky [12], is a Liapunov function for the dynamics, i.e. it is monotonically decreasing. The attractors are therefore the minima of $F(m)$: patterns or mixture dates. Only a small number of patterns is investigated, otherwise "spin–glass states" have to be considered as well [11, 12].

Note that this type of dynamics can be formulated strictly in terms of the overlaps. Consequently different initial states having the same overlaps with the patterns evolve toward the same final state. This is of advantage for the analysis of the model [11, 12] but can be a disadvantage for the performance of the network as demonstrated in the preceding example.

4.2 Generating potential dynamics

At first sight the maximum field dynamics appears to contradict the requirement that any computing is done locally on each neuron.

At this point it is helpful to investigate what neurons actually do, at least on a superficial level [31, 8, 32]. The primary variable describing the state of a neuron is its "generating potential" v_i. In the absence of any input from other neurons it is kept at a level $v_0 \simeq -80mV$, the resting potential, by an interplay of ionic pumps and ohmic shunt currents. Neurotransmitters released from synapses of firing neurons open Na^+ or Cl^- channels which increase or decrease the potential. Their efficiency is denoted by J_{ij} with $J_{ij} > 0$ for excitatory synapses (Na^+ channels) and $J_{ij} < 0$ for inhibitory (Cl^-) synapses. If the potential raises above some threshold $v_{\text{th}} \simeq -50\ mV$ the neuron starts to fire at a rate $\sigma(h)$, given e.g. by (3.7) or (3.8) with h_i replaced by $u_i = v_i - v_{\text{th}}$. This is modelled by the following set of equations

$$\begin{aligned} \tau_0 \dot{u}_i &= -u_i + h_i \\ h_i &= \sum_j J_{ij} \sigma_j \end{aligned} \tag{4.5}$$

$$\begin{aligned} \sigma_i &= \text{th } \beta u_i \qquad \text{or} \\ \sigma_i &= \text{sign } u_i. \end{aligned} \tag{4.6}$$

The relaxation time τ_0 is of the order of a few milliseconds. This description neglects time delays due to the propagation of signals along the dendrites and the axon and treats individual firings only as an average. This might actually be problematic since the delays on the dendrites and the refractory periods between consecutive firings are of the same order as τ_0.

Assume the initial state is given by

$$u_i(0) = \sigma_i(0) = \pm 1 \ . \tag{4.7}$$

Adopting the second choice of (4.6), i.e. for McCulloch–Pitts neurons, the potentials start to relax toward $h_i(0)$, given by (4.5) until the first one crosses zero, let's say at time t_1. This is the one, or the group, for which $-\sigma_i(0)h_i(0)$ is maximum. At t_1 the first group of neurons switches and the local fields h_i adjust accordingly to new values $h_i(t_1)$. For $t > t_1$, the potentials relax toward $h_i(t_1)$ until the potentials of the next group cross zero at time t_2. This proceeds until for all neurons $\sigma_i(t)h_i(t)$ is positive. The sequence of switching neurons is in many cases which we have tried [25] identical to the one prescribed by "maximum field" dynamics, but not in all. The basins of attraction of the retrieval states are larger than those of the Glauber dynamics but slightly smaller than those of the maximum field dynamics.

5. Retrieval from a Reset State

In the example of section 3.4 two inputs, A and B, have been investigated. They both have identical overlaps with the patterns. Input A evolves under the "maximum field" or the "generating potential" dynamics toward pattern 1 having the largest initial overlap, whereas B evolves toward a mixture state. There is an obvious difference between A and B: input A has 9 active neurons, 8 of which are also active in pattern 1, and it contains a single "mistake". Input B has 25 active neurons, 16 correct ones and 9 mistakes. Apparently the performance is better for completing a pattern than for correcting mistakes. This interpretation is, however, based on an asymmetric rating. As a mistake we have counted neurons which are active in the input state and inactive in the pattern with largest overlap, but not the opposite. An equal rating of both yields the overlap and does not distinguish the two initial states.

5.1 Subpatterns

In order to analyze this striking behaviour the concept of sublattice magnetizations, introduced by van Hemmen *et al.* [19] in the context of nonlinear learning rules, is useful in a slightly modified form. The rows of the $N * p$ matrix X, formed by the patterns, are binary words of length p with entries ± 1. The subset of indices belonging to a given word or its negative, i.e. the word with 1 replaced by -1 and vice versa, determines the corresponding subpattern. The number of subpatterns is $n \leq 2^{p-1}$. The maximum number is typically reached for $p \ll \log_2 N$.

Assume the subpattern a contains N_a indices. N_a^+ of them are represented by the corresponding word and $N_a^- = N_a - N_a^+$ by its negative. An $N * n$ matrix T is defined by its elements $t_i^a = 1$ if i belongs to subpattern a and is represented by the word, $t_i^a = -1$ if i is represented by its negative and $t_i^a = 0$ if i does not belong to a. The $n * N$ matrix $\hat{T}$ with elements $\hat{t}_a^i = \pm 1/N_a^\pm$ if $t_i^a = \pm 1$ and $\hat{t}_a^i = 0$ if $t_i^a = 0$, fulfills

$$\hat{T}T = I_n \tag{5.1}$$

where I_n is the n–dimensional identity. We allow for nonorthogonal patterns but the $N_a^{\pm}$ have to be nonvanishing.

The transformation

$$\varphi = \hat{T}\sigma \tag{5.2}$$

defines n subpattern variables $-1 \leq \varphi_a \leq 1$. The transformed patterns form an $n * p$ matrix

$$Z = \hat{T}X. \tag{5.3}$$

Let $\hat{Z}$ be the pseudo–inverse [7] of Z. Then the optimal choice for the synaptic strengths is

$$J = T\ K\ \hat{T} \ \text{ and } \ K = Z\hat{Z}. \tag{5.4}$$

If n is maximum, i.e. $n = 2^{p-1}$, it can be shown [25] that $\hat{Z}$ is $1/n$ times the adjoint of Z and K has the form given in the Hopfield model

$$K_{ab} = \frac{1}{n}\sum_{\mu} \xi_a^{\mu}\xi_b^{\mu} \tag{5.5}$$

where $\xi_a^{\mu} = \pm 1$ are the components of Z.

5.2 Generating potential dynamics

The study of the action of the dynamics with initial values given by (4.7) is simplified if we introduce for each subpattern potentials

$$u_a^s(t) = t_a^i u_i(t) \qquad s = \sigma_i(0)t_i^a = \pm 1 \tag{5.6}$$

having initial values $u_a^s(0) = s$. The resulting equations of motion are

$$\dot{u}_a^s(t) = -u_a^s(t) + k_a(t) \tag{5.7}$$

with fields

$$k_a(t) = \sum_b K_{ab}\varphi_b(t). \tag{5.8}$$

The values of the subpattern variables $\varphi_a(t)$ are found from eqs. (4.6) and (5.2).

The temporal development of the $u_a^s(t)$, $\varphi_a(t)$ and the resulting overlaps $m_\mu(t)$ are shown in Fig. 5 for the example of section 3.4 and the initial state A. The successsive switching and the refinement of retrieval can be read off from the development of $m_\mu(t)$. Note that at each switching a new subpattern is completed, i.e. the corresponding variable φ_a reaches its extremal value, +1 in the example. Figure 6 shows the same for the initial state B which evolves toward the mixture state C.

5.3 The reset state

The definition of T and $\hat{T}$ singles out a particular set of states ρ such that

$$\hat{T}\rho = 0. \tag{5.9}$$

Among those is the "empty" state, $\rho_i = -1$ for all i, i.e. the state where all neurons are inactive. For reasons discussed later this state is called the "reset state". The initial values

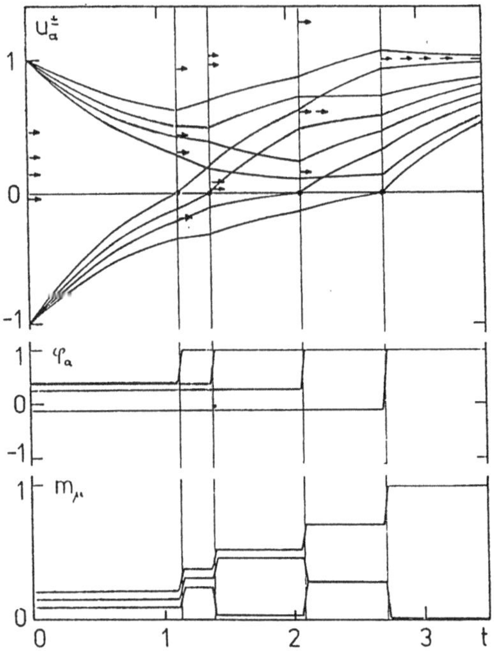

Fig. 5

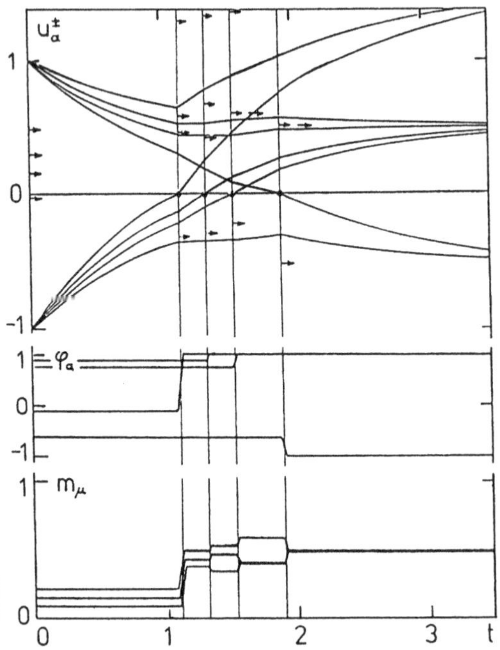

Fig. 6

Fig. 5: Temporal development of $u_a^\pm(t)$, $\varphi_a(t)$ and $m_\mu(t)$ for the input A under generating potential dynamics. The local fields k_a are indicated by arrows.

Fig. 6: Same as Fig. 5 for input B.

of the subpattern variables measure how close the initial state is to this reset state. For the example they are 6/16, 6/16, 4/16 and -2/16 for A and -2/16, 14/16, 12/16 and -10/16 for B. State A is closer to the reset state than B. The asymmetric rating mentioned above is in reference to this reset state. This interpretation makes sense if we can find some reasoning why this reset state is distinct from other states fulfilling (5.9) or, if we allow for different definitions of T and $\hat{T}$, from an even larger set of states. The example suggests that states close to the reset state evolve toward patterns and that some minimal initial distance is required if the state evolves toward a mixture state.

It can be shown [25], for instance, that under maximum field dynamics an initial state can evolve toward a symmetric mixture of three states only if $|\varphi_a(0)| > 1/3$ holds for at least three subpatterns. This is a necessary but not sufficient condition. This suggests the structure of the basins of attractions of patterns and mixture states sketched in Fig. 7. This is in contrast to the Hopfield or Little dynamics which leads to basins of attraction sketched also in Fig. 7, for $\beta \to \infty$.

5.4 Input via external fields

This suggests the following scenario for the operation of the brain. The net describes a certain area. This area receives its input from synapses attached to neurons outside this

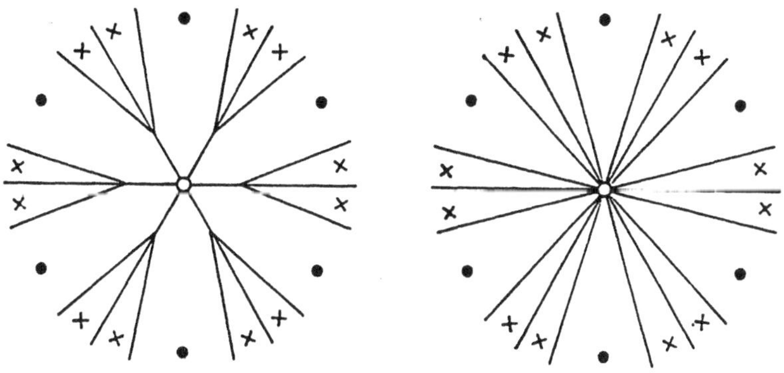

Fig. 7:
Schematic plot of the basins of attraction for patterns (black dots) and mixture states (crosses). The origin ($\varphi_a = 0$) is marked by an open circle. Left: maximum field, right: Glauber mean-field dynamics.

area, neurons belonging to sensory organs or other areas. Their action is modelled by adding external fields h_i^{ex} to the internal fields given by (4.5).

Initially the net is set into its reset state and the generating potentials evolve only under the influence of the external fields. The first switching produces a state of the kind used as input in the previous discussions and the further evolution is, at least qualitatively, as described above. This has beentested in various examples. This kind of operation and the particular choice of the reset state appears not only plausible but is supported by various experiments in which a test person has been asked to perform certain tasks and where the activity has been monitored in certain areas of the brain [33].

6. Outlook

Methods of statistical mechanics and numerical stimulations have produced considerable insight into the functioning of models of neural nets, used as auto–associative memory. Almost exclusively Glauber or synchronous parallel dynamics have been studied for the process of retrieval.

We have seen in an example and from further analysis that dynamics based on generating potentials of neurons or the discrete maximal field dynamics lead to a qualitatively different performance. The basic differences are: i) dynamics is based on a modelling of neurons, ii) retrieval is performed by successive refinement, iii) the existence of a specific reset state is suggested, iv) the performance is better for the task of completing patterns compared to error correction.

Learning and retrieval of hierarchically organized patterns is possibly of great relevance [23]. For such structures the successive refinement is expected to generate automatically a hierarchical retrieval process.

The basic difference in the two types of dynamics can be seen from the equations of motion, eq.(4.1) for Glauber–mean field dynamics and eq.(4.5) for generating potential dynamics. Assuming McCulloch–Pitts neurons, only the signs of the local fields enter for the former, whereas the strongest local field has the strongest influence on the latter.

We have not addressed the important question of learning within Hebb's rule [4] and have assumed that it is done optimally in the sense of Kohonen [7]. Iterative procedures have been proposed by Diederich *et al.* [14], the key appears to be the observation that patterns are distinct from other states, e.g. mixtures, by the fact that their local fields all have the same magnitude. Dream and sleep again might be important in this context [34].

Our investigations have been restricted to small numbers of patterns $p \ll \log_2 N$. It remains to be seen how generating potential dynamics works for $p \sim N$. Simulations are on the way, investigations of the type presented in the work of Amit *et al.* [11] are desirable but probably more difficult because the overlaps m_μ and spin–glass orderparameters are not sufficient to describe the system in mean field theory.

Acknowledgements

Many of the thoughts presented here have originated from discussions with my research students, especially D. Bormann, M. Frick and A. Schmidt. I have also benefited from discussions with L. van Hemmen and R. Kühn, especially on their concept of sublattice magnetizations, and with I. Morgenstern.

References

1. J.J. Hopfield: Proc. Natl. Acad. Sci. USA, **79**, 2554 (1982); ibid: **81**, 3088 (1984)
2. W.S. McCulloch and W.A. Pitts: Bull. Math. Biophys. **5**, 115 (1943)
3. W.A. Little: Math. Biosci. **19**, 101 (1974); W.A. Little and G.L. Shaw: Math. Biosci. **39**, 281 (1978)
4. D.O. Hebb: The Organization of Behavior (Wiley, N.Y., 1949)
5. K. Steinbuch: Die Lernmatrix, Kybernetik **1**, 36 (1961)
6. L.N. Cooper: In Proceedings of the Nobel Symposium on Collective Properties of Physical Systems, ed. by B. Lundqvist and S. Lundqvist (Academic, N.Y., 1973)
7. T. Kohonen: Associative Memory (Springer, 1977); Selforganization and Associative Memory (Springer, 1984)
8. S. Grossberg: Proc. Natl. Acad. Sci. USA, **60**, 758 (1968); Studies of Mind and Brain (Reidel Publ. Co. Dordrecht, 1982)
9. R.J. Glauber: J. Math. Phys. **4**, 294 (1963)
10. See e.g. I. Morgenstern and L. van Hemmen, eds.: Heidelberg Colloquium on Glassy Dynamics and Optimization (Springer, 1987)

11. D.J. Amit, H. Gutfreund and H. Sompolinsky: Phys. Rev. **A32**, 1007 (1985); Ann. Phys. N.Y., **173**, 30 (1987); Phys. Rev.**A35**, 2293 (1987)

12. I. Kanter and H. Sompolinsky: Phys.Rev.**A35**, 380 (1987)

13. L. Personnaz, I.Guyon and G. Dreyfus: J. Physique Lett.**46** L359 (1985); Phys. Rev. **A34**, 4217 (1986)

14. S. Diederich, M. Opper, R.D. Henkel and W. Kinzel (to be published); S. Diederich and M. Opper, Phys. Rev. Lett. **58**, 949 (1987)

15. H. Ritter and K. Schulten: Biol. Cybern.**54**, 99 (1986); H. Sompolinsky and I. Kanter (to be published)

16. H. Sompolinsky (to be published)

17. B. Derrida, E. Gardner and A. Zippelius (to be published)

18. W. Kinzel: Z. Physik **B60**, 205 (1985)

19. J.L. van Hemmen and R. Kühn: Phys. Rev. Lett.**57**, 913 (1986); J.L. van Hemmen, D. Grensing, A, Huber and R. Kühn (to be published); J.L. van Hemmen (to be published)

20. I. Morgenstern (to be published)

21. J.P. Nadal, G. Toulouse, J.P. Changeux and S. Dehaen: Europhys. Lett. **1**, 535 (1986)

22. G. Parisi: J. Phys. **A19**, L 617 (1986)

23. N. Parga and M.A. Virasoro: J. Physique **47**, 1857 (1986); H. Gutfreund (to be published)

24. J.S. Denker, ed.: Proc. Conf. Neural Network for Computing (Am. Inst. of Phys. 1986)

25. H. Horner, D. Bormann, M. Frick and A. Schmidt (in preparation)

26. R. Rammal, G. Toulouse and M.A. Virasoro: Rev. mod. Phys. **58**, 765 (1986); K. Binder and A.P. Young: Rev. mod. Phys.**58**, 801 (1986)

27. D. Sherrington and S. Kirkpatrick: Phys. Rev. Lett.**35**, 1792 (1975)

28. H. Horner, Z. Physik **B66**, 175 (1987); H. Horner and H. Kinzelbach (to be published)

29. G. Palm: In Brain Theory, ed. by G. Palm and A. Aertsen (Springer 1986)

30. G. Toulouse, S. Dehaen and J.P. Changeux: Proc. Nat. Acad. Sci, USA, **83**, 1695 (1986)

31. High-school text books on biology

32. J.J. Hopfield and D.W. Tank: Science **233**, 625 (1986)

33. P.E. Roland and L. Friberg: J. Neurophysiol.**53**, 1219 (1985)

34. F.C. Crick and G. Mitchison: Nature **304**, 111 (1983); J.J. Hopfield, D.I. Feinstein and R.G. Palmer: Nature **304**, 158 (1983)

Mean-Field Theory of Spin Glasses and Neural Networks with Finite Coordination Number

I. Kanter

Department of Physics, Bar-Ilan University, Ramat-Gan 52100, Israel

ABSTRACT

The mean-field theory of dilute spin glasses and neural networks is studied in the limit where the average coordination number is finite (i.e., the average number of neighbors connected to each site). The zero-temperature phase diagram is calculated. Comparison between the properties of dilute neural networks and fully connected nets is presented. The relationship between the different phases and the percolation transition is discussed.

1. INTRODUCTION

Spin glasses and neural networks are examples of interacting systems that undergo ordering at low temperatures despite their randomness and frustration. The theoretical understanding of the formation of long-range order in random frustrated systems is based almost entirely on the results of mean-field theory (MFT). Usually MFT assumes that the systems possess a very high degree of connectivity. Specifically, if the number of degrees of freedom (spins or neurons) is N, then it is hypothesized that each one is connected to the rest N-1 degrees of freedom. This assumption of full connectivity is an idealization which is almost never realized in physical or biological systems. Typically, the average coordination number, i.e., the average number of bonds per degree of freedom, is much smaller than N. It is thus important to study the role of the connectivity in shaping the cooperative dynamical behavior of spin glasses and neural networks.

In this paper, the MFT of dilute spin glasses and neural networks is studied. Two cases must be distinguished. If, after dilution of the bonds, the average number of bonds per spin or neuron remains of O(N), the dilution does not affect the behavior of the system.[1] If, however, the average coordination number is finite, some new physics is expected to emerge.[2-3] In particular, there will be an interesting interplay between the statistical-mechanical frustration and the geometric-connectivity fluctuations. Besides the relevance to the properties of spin glasses and neural networks, the theory of dilute spin glasses has important applications in graph optimization problems.[4-5] Some of these problems, such as graph partitioning and graph coloring, can be mapped into random, frustrated Ising models with highly diluted infinite-range interactions.

The paper is organized as follows: In section 2, a review is given of the MFT of a system with finite coordination number, which has recently been studied by Kanter and Sompolinsky.[2] The phase diagram and the properties of the low-T phases are discussed. In Section 3, the MFT of a dilute Hopfield model with finite coordination number is solved and its properties are discussed and compared to the undiluted case. Finally, a discussion is given in section 4.

2. THE MODEL

We consider an Ising system described by the Hamiltonian

$$H = \frac{1}{2} \sum_{i \neq j} J_{ij} S_i S_j , \tag{1}$$

where $S_i = \pm 1$ (i=1,...N) and the J_{ij}'s are infinite-ranged random interactions. Their probability distribution is

$$P(J_{ij}) = (1-c/N)\delta(J_{ij}) + (c/N) f(J_{ij}) . \tag{2}$$

This describes a network of bonds which is highly diluted: The average coordination number of each spin or neuron is c which is taken to be of order unity. The distribution of the surviving bonds is given by $f(J_{ij})$ which is normalized to unity. Because the average number of bonds is cN/2 (and not $N^2/2$), the scale of J_{ij} must be of order unity to obtain the appropriate thermodynamic limit.

Despite the dilution, the interactions J_{ij} have infinite range, i.e., each pair of spins in the system has the same <u>probability</u> of being coupled.

Since there are no length scales in the problem, a mean-field theory is expected to give an exact description of the system in the thermodynamic limit ($N \to \infty$). Indeed, using the replica method,[6] one can show that when the bonds are independent variables the average free energy per spin at temperature $T = \beta^{-1}$ can be expressed[7] as

$$\beta f = \frac{c}{n} \left\{ \frac{a_1}{2} \sum_{\alpha} Q_{\alpha}^2 + \frac{a_2}{2} \sum_{\alpha<\beta} Q_{\alpha\beta}^2 + \frac{a_3}{2} \sum_{\alpha<\beta<\gamma} Q_{\alpha\beta\gamma}^2 + \ldots \right\}$$

$$- \frac{1}{n} \ln \mathop{\mathrm{Tr}}_{\{S^{\alpha}\}} \exp(-\beta \bar{H}) - \frac{c}{n} \ln \int dJ f(J) \cosh^n(\beta J) , \tag{3}$$

$$\beta \bar{H} = -c \left[a_1 \sum_{\alpha} Q_{\alpha} S^{\alpha} + a_2 \sum_{\alpha<\beta} Q_{\alpha\beta} S^{\alpha} S^{\beta} + a_3 \sum_{\alpha<\beta<\gamma} Q_{\alpha\beta\gamma} S^{\alpha} S^{\beta} S^{\gamma} + \ldots \right] . \tag{4}$$

The indices $\alpha, \beta, \gamma, \ldots$ run from 1 to n. The variables S^{α} represent n spins <u>at the same site</u>. The constants a_k are given by

$$a_k = \int_{-\infty}^{\infty} dJ f(J) \tanh^k(\beta J) \tag{5}$$

The physical free energy is derived by minimizing with respect to $(Q_\alpha, Q_{\alpha\beta}, Q_{\alpha\beta\gamma}, \ldots)$ and taking the limit $n \to 0$.

The new feature of Eq. (3) is the appearance of a large number of order parameters, whereas in the MFT of undiluted systems, only Q_α and $Q_{\alpha\beta}$ appear. The present model in the case of spin glasses has been studied in Ref. 7 near the transition temperatures. There, to leading order, one can neglect all but a few of the order parameters which results, not surprisingly, in behavior qualitatively similar to that of the Sherrington-Kirkpatrick (SK) model.[6] Here we focus on the low temperature regime where all order parameters are of the same magnitude. This problem has also been studied recently in refs. 8 and 9.

In this paper we solved the problem within the framework of replica-symmetric theory. The order parameters are assumed to be independent of the replica indices, i.e., $Q_\alpha = Q_1, Q_{\alpha\beta} = Q_2$, $Q_{\alpha\beta\gamma} = Q_3$, etc., for all replica indices. The quantities Q_k are simply the moments of the local magnetizations, $Q_k = \langle\langle m_i^K \rangle\rangle$ where $m_i \equiv \langle S_i \rangle_T$, where $\langle \ldots \rangle_T$ is the thermal average, and $\langle\langle \ldots \rangle\rangle$ stands for an average over the J_{ij}. On the assumption of this structure, the limit $n \to 0$ of Eq. (3) can be taken explicitly, yielding a free energy which is a function of all order parameters Q_k. Instead of dealing directly with an infinite number of order parameters, it is most useful to consider the probability distribution of the local fields defined by $h_i \equiv \tanh^{-1}\langle S_i \rangle_T$. Note that h_i is not equivalent to the <u>exchange</u> field $\sum_j J_{ij} m_j$. As $T \to 0$, $T|h_i|$ is the minimum energy required to change the i-th spin from its ground state by an arbitrary excitation which involves the flipping of a finite number of spins. The free energy can be expressed as a functional of the averaged, local field distribution P(h). Extremizing this free energy, we have derived the following self-consistent equation for the "order function" P(h):

$$P(h) = e^{-c} \int_{-\infty}^{\infty} \frac{dy}{2\pi} \exp\left[-iyh + c \int_{-\infty}^{\infty} dJ \bar{f}(J) \int_{-\infty}^{\infty} P(x)dx \, \exp\left\{iy \tanh^{-1}(\tanh(\beta J)\tanh x)\right\}\right] , \quad (6)$$

where $\bar{f}(J)$ is exactly equivalent to f(J) in the case of independent bonds (SG case), and in the case of dependent bonds (Hopfield model), the relation between $\bar{f}(J)$ and f(J) will be given in Eq. (19). Near the transition temperature the local fields are small and hence one can expand the exponent in Eq. (6) in powers of m(x)=tanhx. This leads to self-consistent equations for the lowest moments of m.[7] For general T, and arbitrary bond distribution f(J), Eq. (6) can be solved numerically. Here, we specialize to two particularly simple cases of discrete bond distribution in the limit of $T \to 0$.

3. DILUTE SG

The first case to be considered is that of ±1 bonds distribution

$$f(J) = a\delta(J-1)+(1-a)\delta(J+1) \quad . \quad (7)$$

Since in this case the excitation energies are integers, P(h) must have at zero T, the following form:

$$P(h) = (1-Q)\delta(h)+\sum_{\ell=1}^{\infty}[P^{+}_{\ell}\delta(h-\beta\ell)+P^{-}_{\ell}\delta(h+\beta\ell)] \quad . \tag{8}$$

Using the definitions $P^{\pm} = \sum_{\ell} P_{\ell}^{\pm}$, it is evident that $P^{+}+P^{-} = Q$, $P^{+}-P^{-} = 1$ where Q is the total fraction of frozen spins, and m is the net magnetization (per spin) of the frozen spins. Taking the $T \to 0$ limit of Eq. (6) we note that $\lim_{T\to 0}|\tanh^{-1}[\tanh(\beta J)\tanh x]|$ is equal to $\beta|J|$ if $|x|\geq\beta J$, and to $|x|$ if $|x|\leq\beta(J)$. Substituting Eq. (8) into the right-hand side of Eq. (6), one finds

$$P(h) = e^{-c}\int_{-\infty}^{\infty}\frac{dy}{2\pi}\exp[-iyh+cQ(x_{+}e^{+iy\beta}+x_{-}e^{-iy\beta})] \tag{9}$$

where

$$x_{\pm} = \tfrac{1}{2} \pm (m/Q)(a-\tfrac{1}{2}) \tag{10}$$

Expanding the integrand of Eq. (9), after some algebra one finds two self-consistent equations for m and Q

$$1-Q = e^{-cQ}I_{0}(2cQ(x_{+}x_{-})^{\frac{1}{2}}) \tag{11}$$

$$m = cQe^{-cQ}\int_{x_{-}}^{x_{+}}dt\{I_{0}(2cQ[t(1-t)]^{\frac{1}{2}}) + [4t(1-t)]^{-\frac{1}{2}}I_{1}(2cQ[t(1-t)]^{\frac{1}{2}})\} \tag{12}$$

where the $I_{\nu}(x)$ are modified Bessel functions.

We now discuss a few consequences:

(I) Dilute ferromagnet: a = 1. In this case $P^{-} = 0$ and Eqs. (11) and (12) reduce to Q = m = P, where P is the order parameter of the infinite-range percolation satisfying the relation[10]

$$1 - P = e^{-cP} \quad . \tag{13}$$

It is non-zero above the percolation threshold c = 1, and approaches unity when $c \to \infty$ as $P \sim 1-e^{-c}$. In addition, Eq. (9) yields the interesting result concerning the structure of the infinite cluster that $P_{\ell}^{+} = (cP)^{\ell}e^{-cP}/\ell!$. Note that by its definition, Eq. (8), P_{ℓ}^{+} is the average concentration of spins that can be disconnected from the infinite cluster by cutting only ℓ bonds. When calculating a dilute ferromagnetic system with magnetic field, one can use similar methods to calculate the distribution of the number of spins that can be disconnected from the infinite cluster by cutting only ℓ bonds.

(II) Spin-glass phase: a < 1. For all a < 1, there is a range of c > 1 where a spin-glass phase exists, characterized by m = 0, Q ≠ 0. In this phase, Eq. (11) reduces to

$$1 - Q = e^{-cQ}I_{0}(cQ) \tag{14}$$

which is non-zero above c = 1 with Q = (4/3)(c-1) near $c \sim 1^{+}$. As $c \to \infty$, Q approaches unity only as a power law, $1-Q \sim (2\pi c)^{-\frac{1}{2}}$. Note that for all c > 1, Q is less than P as shown in the inset of Fig. 1.

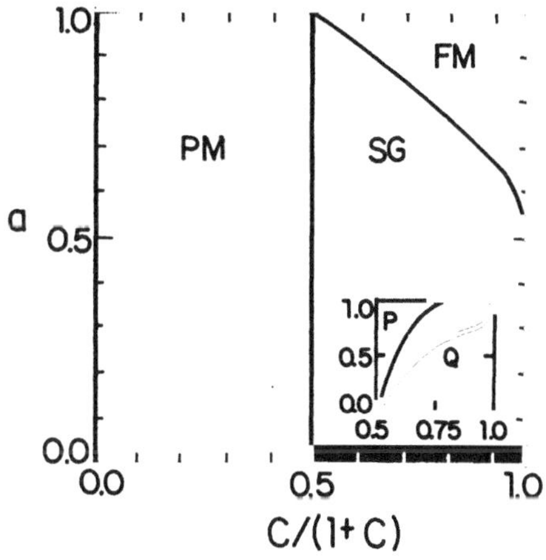

Fig. 1:
The zero-T phase diagram: a is defined in Eq. (7); PM, FM, and SG denote the paramagnetic, ferromagnetic, and spin glass phases. Inset: The percolation order parameter, P, Eq. (13), and the SG order parameter, Q, Eq. (14), as a function of c/(1+c) in the SG phase

The difference P-Q represents the average concentration of frustrated spins, i.e., the concentration of spins on the infinite cluster which can be flipped at T = 0 by an excitation of zero energy. Another aspect of the frustration is given by the ground state energy, E_{SG}. The quantity $N(E_{SG}+c/2)/2$ is the total number of unsatisfied bonds in the SG phase. Near the transition, $E_{SG}+c/2 \simeq 1/8(c-1)^3$.

(III) Ferromagnetic phase: $\frac{1}{2} < a < 1$. Expanding Eq. (12) for small m, one finds a transition from the SG phase to a ferromagnetic phase which occurs at the critical value of c given by the following equation:

$$1 = (2a-1)ce^{-cQ}[I_0(cQ) + I_1(cQ)] , \tag{15}$$

where Q is the SG order parameter. For $0 \leqslant a \leqslant \frac{1}{2}$, Eq. (15) does not have a solution, implying that when the concentration of negative bonds is higher than the positive bonds, the SG phase exists at T = 0 for all c. For $a \geqslant \frac{1}{2}$, a ferromagnetic phase appears above the critical c and is characterized by $q > m > 0$.

The full phase diagram at T = 0 is shown in Fig. 1. Note that although the undiluted, infinite-range antiferromagnet remains paramagnetic even at zero temperature, the dilute antiferromagnet (the present model with a = 0) freezes into a SG state for all c > 1. Thus, any arbitrarily weak dilution of the infinite-range antiferromagnet pins a large fraction of the frustrated bonds and causes the system to freeze at low T.

Many of the above qualitative results, including the general form of the phase diagram, are valid also for independent bond distributions other than Eq. (7). Perhaps the most important difference is associated with the value of the SG order parameter Q. It is clear that if the bond distribution $f(J_{ij})$ is continuous, the value of Q is exactly the same as that of the percolation order parameter P, Eq. (13). This is indeed expected, since there is zero probability that the local fields on the percolating cluster vanish, and hence all the spins on the infinite cluster are frozen at T = 0.

4. DILUTE HOPFIELD MODEL

The second case to be discussed here is the diluted Hopfield model.[11-12] The energy function of the system is given by Eq. (1). The two states $S_i = \pm 1$ represent the two main levels of activity of the i-th neuron. The surviving bonds J_{ij} are the synaptic efficacies of pairs of neurons. The connections are assumed to be modified by learning in a manner which ensures the dynamic stability of certain configurations. In the "Hebb learning rules" the accumulated effect of learning on the synaptic connection between the pair (i,j) can be represented by

$$J_{ij} = \left(\sum_{\mu=1}^{p} \xi_i^{\mu}\xi_j^{\mu}\right)T_{ij} \quad , \tag{16}$$

where the distribution of T_{ij} is given by

$$P(T_{ij}) = (1 - \frac{C}{N})\delta\,(T_{ij}) + \frac{C}{N}\,\delta(T_{ij} - 1)\ . \tag{17}$$

The p patterns $\{\xi_i^{\mu}\}$ constitute the embedded memories. They are assumed to be random with equal probabilities for $\xi_i^{\mu} = \pm 1$. Once again, the scale of J_{ij} must be of order unity to achieve the appropriate thermodynamic limit. Here, we focus mainly on the spin-glass states and retrieval states which are most important for the retrieval of memory. A retrieval state is a state which has a large overlap with a single pattern. The overlap of each state with corresponding pattern is

$$m = \frac{1}{N}\sum_i \xi_i^{\mu} S_i \tag{18}$$

A spin glass state is a stable state in which all overlaps are of $O(1/\sqrt{N})$. Extremizing the free energy, one finds that the equation for the "order function" P(h) is given by Eq. (6) where the distribution of the strength of the bonds are of the type of Hopfield model[11-12] after the Mattis transformation

$$\mathcal{T}(J) = 2^{-p+1} \sum_{\ell = -p+1,-p+3,\ldots p-1} \delta(J-\ell-1) \begin{bmatrix} p-1 \\ \frac{p-\ell-1}{2} \end{bmatrix} . \tag{19}$$

It is important to note that in the Hopfield model, bonds which form a closed loop are correlated. Equation (19) shows that the correlations among bonds do not affect the form of the self-consistent equations of the "order function" P(h). The neglect influence of the correlations among bonds on the order function is due to the fact that in highly diluted systems, the length of a typical loop is of order logN, and therefore correlation can be neglected.[13] In the limit $T \to 0$, the excitation energies are integers, and P(h) must have the following form:

$$P(h) = (1-Q)\delta(h) + \sum_{\ell=1}^{p-1}[P_\ell^+\delta(h-\beta\ell) + P_\ell^-(h+\beta\ell)] + P_p^+\theta(h-\beta p)$$

$$+ P_p^-\,\theta(-h-\beta p) \quad , \tag{20}$$

where $\theta(x)$ equals zero for $x < 0$ and unity for $x > 0$. Substituting eq. (20) into the self-consistent equation for the "order function" P(h), Eq. (6), gives after some algebra 2p-1 self-consistent equations. The numerical solutions of the equations for p equal to 3 and 15 are given in Fig. 2 where m(c) is plotted vs. c/p.

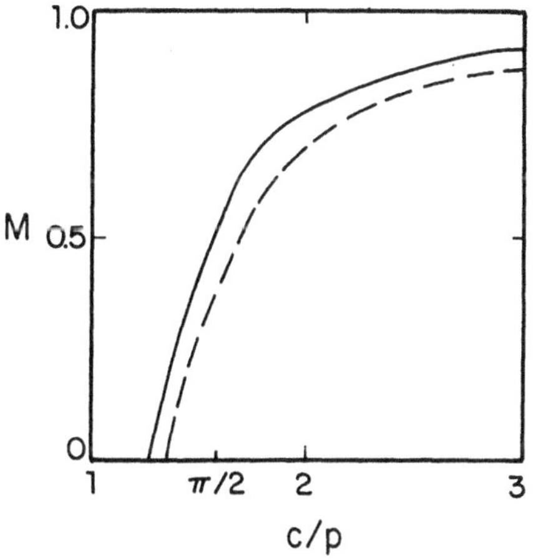

Fig. 2: The magnetization as a function of p is plotted vs. c/p. The full line stands for p = 3 and the dashed line stands for p = 15. $\pi/2$ is the critical value of c when p and $c \to \infty$

The critical value of c above which the retrieval state first appears is ~3.8 for p = 3 and ~20 for p = 15. The diluted Hopfield model has a SG phase for the range $1 < c < c^*(p)$. At the critical value $c^*(p)$, the system undergoes a second order phase transition as a function of c. Above the critical $c^*(p)$, there are no SG states, which might indicate that the basin of attraction is large.

In the limit c and $p \to \infty$ (after $N \to \infty$), one finds the following result

$$m = \frac{2}{\sqrt{\pi}} \int_0^{m/\sqrt{2\alpha}} e^{-y^2} dy \quad , \tag{21}$$

where $\alpha = p/c$. The critical value of α in this limit is $\alpha = 2/\pi$ and the transition is second order $m \propto (2/\pi-\alpha)^{1/2}$. It follows from Eq. (21) that for m = 0.97 (which is the overlap at maximum capacity 0.14N, when c = N-1) α = 0.2, which implies a larger capacity per bond than in a fully connected Hopfield model. Some of the above results for the symmetric diluted Hopfield model are similar to that of the diluted asymmetric model.[14] In particular, the magnetization, Eq. (21), and the critical value of c are exactly the same as in the asymmetric case in the limit of p and $c \to \infty$.

Simulations have been carried out for systems with up to 2000 spins in the range $3 \leqslant p \leqslant 11$. Without cooling the system and with the initial conditions $S_i = \xi_i^\mu (i = 1,2,\ldots N)$, there is a high remanent magnetization for all values of c. The remanent magnetization is partially due to finite clusters and trees on the infinite cluster. Cooling the system from high temperatures, we find a few percent deviation from the theoretical curve of m as a function of c.

Another difference between the diluted and the undiluted Hopfield model is the basin of attraction and the fast convergence to the retrieval state. In the undiluted case, the convergence at zero temperature to the retrieval state is very fast (a few Monte Carlo steps per spin), but the basin of attraction is limited by the spurious states. In the diluted case, simulations show stable states at zero temperature that are different from the retrieval states, although the spin-glass phase does not exist in this region. This behavior is the result of spins with a finite barrier to spin-flip, which prevents a finite fraction of spins from being aligned parallel to the pattern. Approaching zero temperature by annealing, the overlap with the retrieval state after up to several hundreds of Monte Carlo steps is close to the theoretical result. The number of Monte Carlo steps to reach this result depends on c,p and the rate of cooling. This result holds also for initial states close to random configurations, which might indicate that the basin of attraction is large. The slow convergence to the pattern is due to slow propagation of information in highly diluted systems as a result of loops of O(logN).

5. DISCUSSION

We have used mean-field theory to obtain the phase diagrams for spin glasses and neural networks with finite coordination number. The percolation transition, the geometric structure of the infinite cluster and the mass of the finite clusters have a strong influence on the behavior of such systems. The temperature also has a strong effect on the dynamics of the system compared to an undiluted system. The temperature affects the remanent magnetization and smooths significantly the basin of attraction. The existence of regions without SG state seems to be attractive for associative memory systems. The interesting question of the influence of large loops on the dynamics of the system is currently under study. The capacity per bond in the highly diluted system seems to be higher than in the undiluted case. Nevertheless, this is not the only criterion for the performance of an associative memory system. The preference of a diluted system or an equivalent undiluted system with more synapses and fewer neurons is still an interesting open question.

ACKNOWLEDGEMENTS

This research has been carried out in collaboration with Professor H. Sompolinsky. I also acknowledge illuminating discussions with Professors D.S. Fisher, D.J. Gross and D. Huse, and thank Professor Nathan Wiser for a critical reading of the manuscript. This research is supported in part by a grant from the U.S.-Israel Binational Science Foundation.

REFERENCES

1. H. Sompolinsky, Phys. Rev. A 34, 2571 (1986).
2. I. Kanter and H. Sompolinsky, Phys. Rev. Lett. 58, 164 (1987).
3. I. Kanter and H. Sompolinsky, preprint.
4. Y. Fu and P.W. Anderson, J. Phys. A 19, 1605 (1986).
5. I. Kanter and H. Sompolinsky, to be published.
6. S. Kirkpatrick and D. Sherrington, Phys. Rev. B 7, 4384 (1978).
7. L. Viana and A.J. Bray, J. Phys. C 18, 3037 (1985).
8. M. Mezard and G. Parisi, preprint.
9. C. De Dominicis and P. Mottishaw, preprint.
10. P. Erdos and A. Reyni, in The Art of Counting, edited by J. Spencer (MIT Press, Cambridge, MA, 1973).
11. J.J. Hopfield, Proc. Nat. Acad. Sci. USA 79, 2554 (1982).
12. D.J. Amit, H. Gutfreund and H. Sompolinsky, Phys. Rev. A 32, 1007 (1985); Phys. Rev. Lett. 55, 1530 (1985).
13. I am grateful to D.S. Fisher for drawing my attention to this point.
14. B. Derrida, E. Gardner and A. Zippelius, preprint.

Neural Networks for Associative Memory Design

L. Personnaz, I. Guyon, and G. Dreyfus

Ecole Supérieure de Physique et de Chimie Industrielles
de la Ville de Paris, Laboratoire d'Electronique,
10, rue Vauquelin, F-75005 Paris, France

1. INTRODUCTION

The recent wave of interest in cellular structures known as neural networks is due, in large part, to the advent of a new model [1] which turned out to be amenable to analytical results with the tools of statistical mechanics [2]. In a surprisingly short period of time, one has gone a very long way from the initial model ; in the present volume, H. Gutfreund presents a review of the basic concepts and of the most recent developments in this very fast-growing field. Apart from the theoretical interest involved in modelling the brain - or at least some functions of the brain - there is also a considerable interest from the point of view of applications to data processing. The present paper will focus essentially on the latter aspect of artificial neural networks. Associative memory is the basic function that can be performed by these systems ; therefore, we shall present a general discussion of the concepts related to associative memory, applied to pattern recognition and error correction ; various illustrative examples will be shown, and we shall discuss the basic issues in this context. We shall also mention recent developments in the storage and retrieval of sequences of pieces of information.

2. NEURAL NETWORKS AND ASSOCIATIVE MEMORIES

This section is devoted to a general presentation both of the neural networks which will be considered in this paper, and of the concept of associative memory. We shall first present the structure and dynamics of the networks, and explain the importance of the existence of attractors for these structures ; the second section will deal with some basic concepts related to associative memories.

2.1 The structure and dynamics of neural networks

The neural networks considered here are assemblies of McCulloch-Pitts binary formal neurons, with the following operation : each neuron computes the weighted sum of its inputs and makes a decision by comparing it to a predetermined threshold; if the sum is larger than the threshold, the neuron goes to (or remains in) the active state ; if the sum is smaller than the threshold, the neuron goes to (or remains in) the inactive state. We denote the state of neuron i by a variable σ_i which takes the values +1 or -1 ; C_{ij} is the weight of the synapse inputting information from neuron j to neuron i ; we shall take all thresholds equal to zero. Such neurons are arranged to form a fully connected network. Therefore, the evolution of the state of neuron i is governed by the following process : denoting by v_i the potential of neuron i,

$$v_i(t) = \sum_j C_{ij}\, \sigma_j(t) \quad \text{, one has}$$

$$\sigma_i\,(t+\tau) = \operatorname{sign} v_i(t) \,,$$

where τ is a characteristic response time of the neuron.

In all the following, we consider neural networks with parallel, synchronous dynamics: all neurons evaluate their potentials and make their decisions simultaneously, with the same response time τ. The state of a network of n neurons is represented by a vector σ whose n components are equal to ± 1. The points in state space that can be occupied by the network are the summits of a hypercube. The dynamics of the network is fully defined by the values of the interaction coefficients C_{ij} ; the matrix C of these coefficients is generally termed the synaptic matrix of the system.

Such neural networks are essentially dynamical systems, used in the following way : the network is forced into some state, then left to evolve freely until it reaches a stable state or a limit cycle which is an attractor. If a stable state is reached, we may consider that the neural network has made some "computation" whose result is the final state. The notion of computation with attractors is central in the recent development of neural networks ; the essential function fulfilled by these networks, in the present state of the art, is that of associative memory [3,4].

2.2 Associative memory

An associative memory is a device which learns associations between source and target information ; once learning has been completed, the memory is able to retrieve the target information when presented with an incomplete or distorted version of the source information. If the source and target information is identical, the system is called an autoassociative memory. Various applications of this function are illustrated on Fig. 1. In case a, the autoassociative memory retrieves the stored information itself ; in cases b and c, the memory retrieves the target information associated with the input.

Given the above-mentioned dynamics of the neural networks, the use of the latter as associative memories is a natural idea : starting from an initial state which is the input information, the network reaches an attractor which is the target information. Beyond this conceptually simple idea, there are a few important facts and problems that are easily overlooked and should therefore be emphasized :

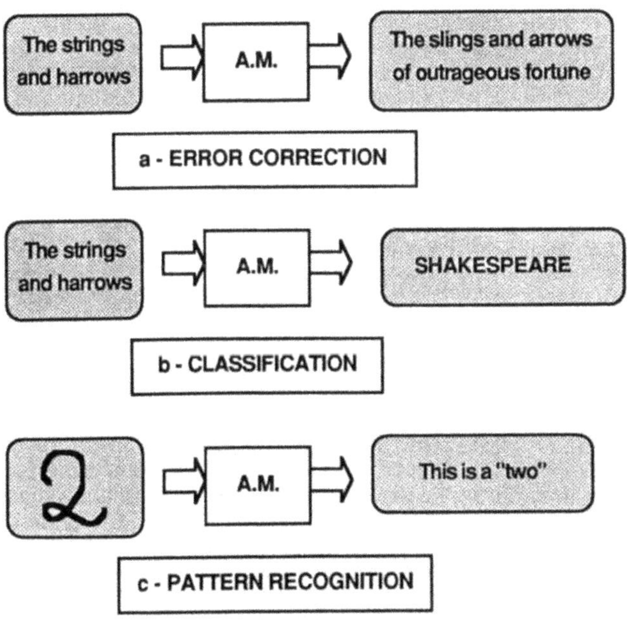

Figure 1 : Various applications of an associative memory

- *Distributed memory and distributed decision* **:** The stored information is distributed among the synaptic coefficients, so that the network is remarkably tolerant to errors in the computation of these values during learning. This is due to the redundancy of the representation of the information : if there are p patterns with n bits each, the total information is np bits, whereas there are n^2 synaptic coefficients, each of which can be stored on approximately $\log_2 p$ bits. Moreover, the network makes decisions at each parallel iteration, so that the final decision is the result of the collective behaviour of the assembly of neurons ; this is to be contrasted to the operation of the conventional linear associative "memories" [5], where some computations are performed, and where subsequent decisions must be made by the user, or by a user-written program. Strictly speaking, a linear system cannot be termed a memory, whereas neural networks, being inherently non-linear, are able to make decisions, distributed both in space (each neuron makes its own decision) and in time (several iterations may be necessary).

- *Information representation and coding* **:** Three issues are important in these respects: *representation, invariance,* and *time* ; we shall discuss these points briefly.

> *Representation* **:** Each neuron, being a binary element, carries a bit of information ; therefore, the state of the neural network at time t is a piece of information coded on n bits. Moreover, since all neurons are identical and since the network is fully connected, all bits have the same information value. Therefore, the items to be stored in the neural network must have a binary representation which complies with this constraint. Numbers must be represented, for instance, by a "thermometer" code.
>
> *Invariance* **:** It is highly desirable, in problems related to vision for instance, to be able to classify an item whatever its size or position. Until now, no neural network has exhibited built-in invariance ; a natural way of obtaining such a property is to provide the neural network with information which is already invariant ; therefore, it is mandatory to find representations which are adapted to neural networks, or, conversely, to design network architectures that can handle invariant representations efficiently.
>
> *Time* **:** It is clear that time is an important parameter in biological memory ; it also has a large impact on the design of artificial neural networks. There are many kinds of information which appear naturally in the form of time sequences, as in music for instance, or which are amenable to time sequences, as in data base management. Therefore, when designing an associative memory with neural networks, it is important to decide from the beginning whether the information to be stored should be treated in the form of time sequences or of fixed points. In a subsequent paragraph we shall show how neural networks can store and retrieve time sequences.

- *Learning* **:** The operation of a neural network as an associative memory implies the storage of transitions between source and target states ; as mentioned above, these transitions are determined by the values of the interaction coefficients C_{ij} ; therefore, learning will be the operation whereby these values are computed ; in the following, various learning rules will be presented or referred to. Essentially, one can distinguish "slow learning", in which the items to be stored are presented over and over again until they are satisfactorily stored, from "one shot" learning, in which each item is presented only once and is stored.We shall mention these types of learning in a subsequent paragraph.

- *Autoassociative memories : State space partitioning, prototype and non-prototype attractors* **:** In an autoassociative memory, information ("prototype states") is stored as fixed points ; since the evolution rule is deterministic, the basins of the attractors (which are determined by the synaptic weights) define a partition of state space. Each attractor has its own basin of attraction, which affects the effectiveness of the associative memory. It should be emphasized that the prototype states are only a subset of the attractors of the network ; there are many other stable attractors, which will be termed non-prototype stable states in the following (and are often termed spurious states in the literature). They play a fundamental part in pattern recognition or error correction, as will be shown later, for the following reason : since the information to be stored and retrieved is essentially binary, the simplest possible way of performing such functions would be to compute the Hamming distances between the unknown pattern and the stored ones, and to pick the closest pattern as the correct information. If there are p stored patterns and n bits of information per pattern, this would require pn Boolean operations, which can be easily implemented in parallel hardware. However, such a system is not self-contained for pattern recognition and error correction, since it cannot detect automatically the existence of ambiguous or irrelevant patterns. The potential use of neural networks in such a context arises from the existence of the non-prototype attractors, which can either be considered as "garbage" states, or be used for increasing the recognition capability of the network ; this will be demonstrated below. Conversely, for data compression purposes, the use of neural networks has been advocated [6] for nearest-neighbour search, with an appropriate learning rule and a hierarchical organization of networks ; when analog neurons are implemented in hardware, this may be an adequate solution.

3. STORING AND RETRIEVING PATTERNS, TAILORING THE BASINS OF ATTRACTION

3.1 The projection rule

As mentioned above, learning consists essentially in computing the interaction coefficients so that a given set of binary information be attractors of the system. It turns out that such a result can be achieved by an infinite number of synaptic matrices ; among those, an especially useful rule, termed the projection rule, has been used for various applications. In the following, we shall omit the technical details of the computation, which will be found in Ref. 4, to focus on its main features, and to illustrate the general ideas stated above.

The basic feature of networks designed with this learning rule is that one can define an energy function, which decreases monotonically during the free evolution of the system until a stable state is reached (i.e., during the information retrieval phase), and whose global minima are the prototype states themselves. This is an important asset of these structures, since both their static and dynamical properties can be investigated very efficiently by the methods of statistical physics. In this context, learning by the projection rule can be described as "digging holes" in state space ; the deepest holes are the prototype states, whereas the non-prototype states have a higher energy (except in some pathological cases which occur with a vanishingly small probability).

Neural networks for *autoassociative* memory, designed with the projection rule, can be used essentially in two simple ways : one can either store one example per class of information, or store several examples per class. As an illustration of the first strategy, we consider automatic error correction of character strings, as in Ref. 4. In such a case, one stores correct sentences - journal titles for instance ; if an erroneous version of one of the sentences is input, one wishes the neural network to evolve until it reaches the correct information. This a typical case in which the simple-minded nearest-neighbour search is not satisfactory, since the system would then give a response even when the input is totally meaningless, or when it is very ambiguous. What would be really desirable is a system which clearly signals that the input is ambiguous or that it has no relation to any of the stored items ; this is exactly what a neural network does. It turns out that the non-prototype stable states, in this case, play the role of "garbage states" (Fig. 2): if the input information is ambiguous, the neural network will converge to one of these states. It is very easy to determine that the state which is reached is not one of the prototypes, since the latter have the smallest possible energy, whereas the garbage states have a higher energy. A similar result can indeed be achieved by a k-nearest neighbour approach, which involves setting decision thresholds; it has been shown in Ref. 7 that the choice of these thresholds is a non-trivial task, whereas the neural

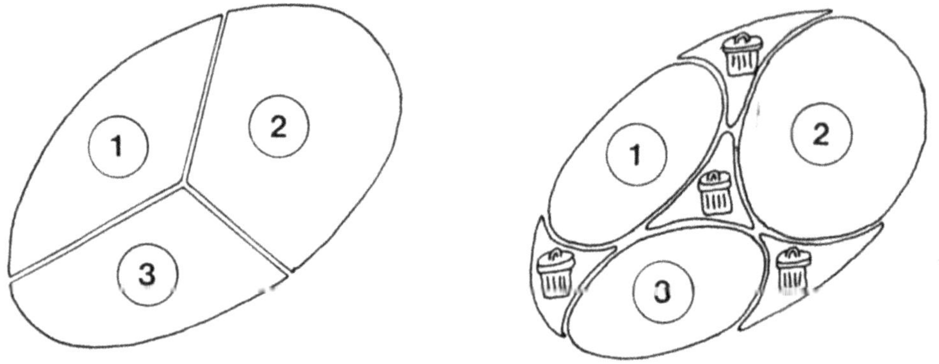

Figure 2 : Left : pictorial representation of the partition of state space for nearest-neighbour search ; right : in a neural network, states which are ambiguous (equidistant or almost equidistant from two, or more, pieces of stored information) are attracted by non-prototype states acting as garbage collectors

networks are able do the task without any adjustable parameters : they are really self-contained machines requiring no additional decision.

The above example showed the use of non-prototype stable states, when one stores one example per class of information ; conversely, one can store several prototypes for each class of information ; a typical example is that of handwritten numeral recognition. In such a case, one stores several examples of each numeral, corresponding to a variety of handwritings and orientations. When presented with an unknown numeral, the system is requested either to recognize the pattern, or to tell the user that it is unable to recognize it. The operation of such a system can be found in Ref. 7. Essentially, this is a "grandmother cell" approach, which can be described briefly as follows : the neural network is made of a number of neurons equal to the number of binary picture elements of the pattern, with ten additional neurons which are used as a position code ; each prototype is stored (using the projection rule) together with its code, which operates as follows (Fig. 3) : all the neurons of the code are inactive, except for one neuron

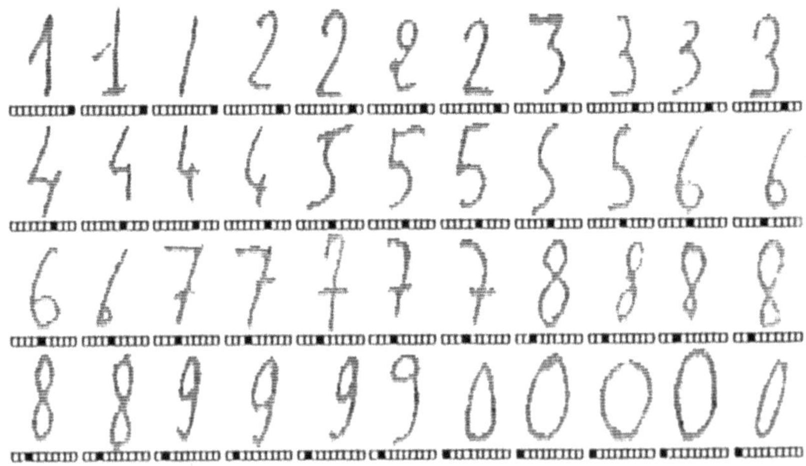

Figure 3 : Prototype states for handwritten character recognition : both the pixels and a code are stored by the projection rule

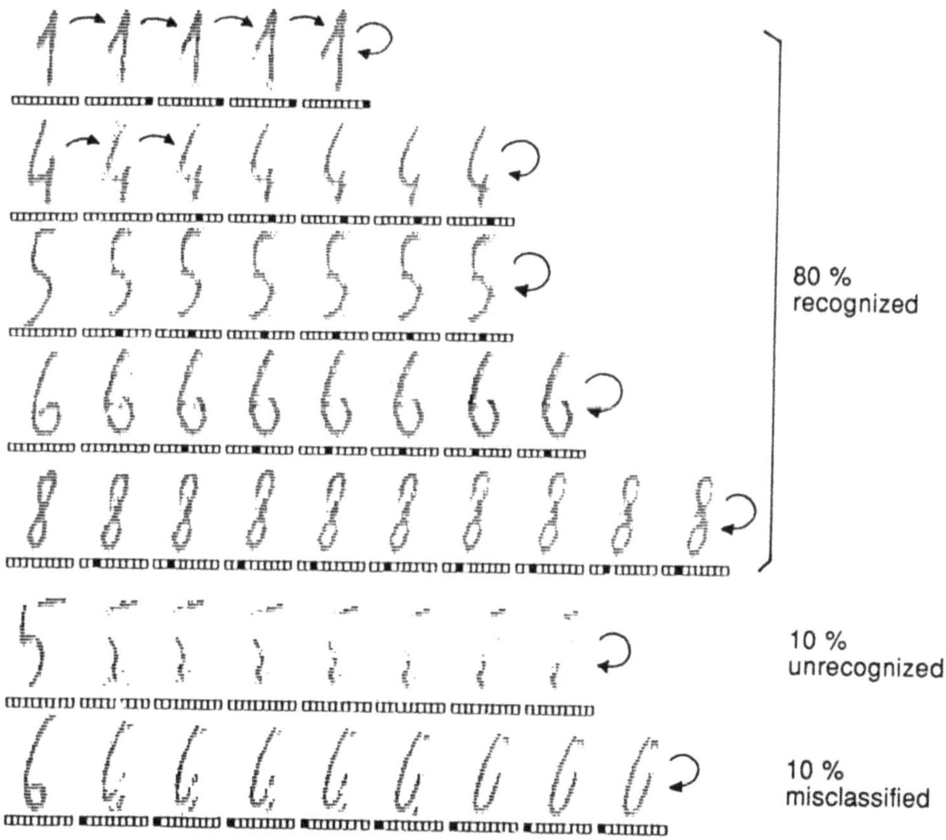

Figure 4 : Information retrieval for handwritten characters : the system retrieves either a prototype state, or a non-prototype stable state having the right code (in both cases, recognition is successful), or a non-prototype stable state with a blank code (no recognition), or a non-prototype stable state with a wrong code (error in recognition)

corresponding to the stored numeral (for instance, the first neuron from the right if the stored pattern is a 1, the second from the right if the pattern is a 2, and so on). In the retrieval phase, the network is set into a state corresponding to the unknown pattern, *with all the neurons of the code inactive*. The network is then left to evolve until it reaches a stable state ; the active neuron in the code field identifies the recognized numeral. The unique property of such a network is the following : as can be seen from Fig. 4, the network will often evolve to a non-prototype stable state, *which looks like a perfectly valid handwritten numeral and exhibits the corresponding code*. This means that, in such a case, the non-prototype stable states do not act simply as garbage states, but play an active and fundamental role in the recognition capabilities of the network. This mechanism is explained in Ref. 7. The results also show that some non-prototype attractor states are used as garbage states ; they are detected through the fact that none of the grandmother cells is active in the final state, or that more than one of them are active simultaneously. This behaviour can be described in terms of an energy landscape (Fig. 5) : each class determines a large basin of attraction, with several minima corresponding to the prototypes of that class and to the non-prototype states which are "mixtures" of the latter ; these basins are separated by energy barriers, at the top of which are non-prototype stable attractors which are mixtures of prototypes corresponding to *different* classes ; they attract the states corresponding to ambiguous

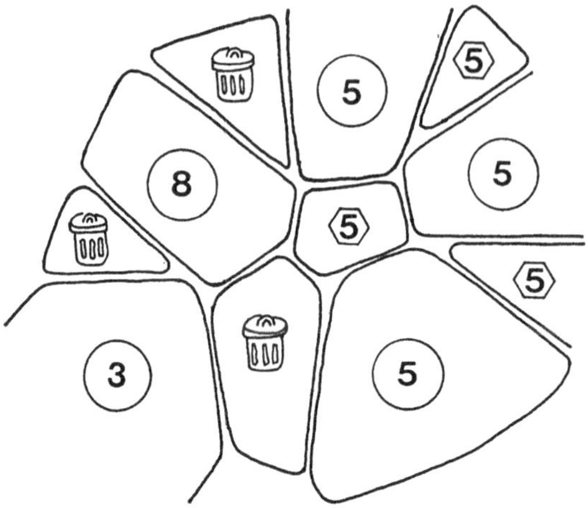

Figure 5 : Pictorial representation of the partition of state space in the previous example: basins whose number are in a circle are the basins of the prototype states ; those with numbers in a hexagon are the basins of non-prototype states having a correct code ; they are smaller and shallower than the previous ones, but still drain an important part of state space, thereby increasing the recognition capabilities of the network

patterns, and are, thus acting as "garbage" states. Therefore, the energy landscape is the image of the structure of the information present in the network.

At this point, the following remark can be made, in conjunction with neurobiology. Assume that the above system is a biological system ; all neurons are strictly identical, and the network is fully connected, so that very little information on its behaviour can be found from anatomical investigations, except for the fact that most cells are connected to "sensory" inputs, possibly through some preprocessing network, whereas some of them (the coding neurons) are not. If the activity of the cells is probed by looking at the responses of the system to the presentation of numerals, one would notice that one given cell is responsive to numerals "1", another cell is responsive to numerals "2", and so on ; therefore, one would be tempted to infer that these are special cells, "trained" to respond to some definite patterns. As can be seen from the above explanations, this would definitely be inaccurate : the coding cells are exactly similar to the other cells. The point is that it is the *collective* behaviour of the network, and the learning rule, which make these cells specifically responsive to certain inputs.

The notion of an energy landscape reflecting the structure of the information has been extended in the following way : since the learning rule creates a meaningful energy landscape, it should be possible to "tailor" or "garden" this landscape in an optimal way, in order, for instance, to accommodate the largest possible amount of information in a network of given size and complexity.

3.2 Imposing transitions and gardening the state space

The basic idea is the following : we have shown that the design of an autoassociative memory may consist in "digging holes" in the energy landscape (store fixed points of the dynamics) ; similarly, it is possible to impose transitions between different states, i.e., to "dig valleys" leading from one state to another, by determining the synaptic coefficients to this effect. This calculation turns out to be just as straightforward as the computation of the synaptic coefficients by the projection rule (see Ref. 4 for more details). This procedure may be used in two different ways : i - imposing transitions from various examples, belonging to the same class, on one single prototype of that class, which is then stored as a fixed point ; ii - imposing transitions from undesirable non-prototype stable states on a prototype state, in order to increase the size of the basin of attraction of this prototype.

An example of the first type of use of the transitions was presented in Ref. 7, for the recognition of handwritten numerals ; the network structure which was described in the previous paragraph was used, but the learning procedure was different : for each numeral, one "archetype" was chosen, as well as several other examples ; learning was performed by imposing transitions from the examples on the archetype, and by storing the latter as a fixed point. This is equivalent to extending the basins of attraction of the archetype towards the various examples of the same class. Such a network turns out to have a recognition capability which is different from that of the previous network (designed with the projection rule), when tested with the same training set. The size of the basins of attraction of the garbage states is reduced ; therefore, the number of misclassifications is increased, while the number of non-recognized characters is decreased.

Suppressing undesired non-prototype stable states can also be achieved by imposing suitable transitions, which has the effect of increasing the size of the basins of attraction [4] ; a systematic approach along these lines has been proposed in Ref. 8, in which a "slow" learning rule is presented. It consists essentially in probing the portion of state space lying within the desired radius of attraction of a given prototype state, and imposing transitions from the probed states on the prototype state. Enormous increases of the radius of attraction have been achieved by using this scheme ; however, it must be pointed out that suppressing all the non-prototype states is not necessarily adequate. Moreover, the efficiency of a neural network which just performs nearest neighbour search is questionable as compared to that of conventional hardware when fully *digital* implementations are considered.

4. STORING AND RETRIEVING TIME SEQUENCES

In the previous section, it was shown that information can be stored in neural networks as fixed points, and that transitions can be imposed in order to alter the energy landscape, i. e., the dynamics of the network. In the present section, we extend the latter idea by imposing sequences of transitions, which is equivalent to storing and retrieving information in the form of time sequences [9]. For instance, a poem can be regarded as a sequence of lines, each line being the state of the network at a given instant of time. The problem which is addressed here is the following : is it possible to compute the synaptic coefficients in such a way that, given a erroneous or incomplete version of one of the lines, the network can retrieve the whole poem ? This problem is the generalization of the idea of imposing transitions since a poem is nothing but a succession of lines (states) A, B, C, ... ; storing a poem consists in imposing a transition from line A to line B, then a transition from line B to line C, and so on.

However, an additional problem appears : a given piece of information may belong to several distinct sequences, or may occur more than once in a given sequence (it is termed a bifurcation point). This is exemplified in Fig. 6, showing an example of a poem whose complicated structure of transitions in state space is apparent.

A simple storage prescription for such sequences can be derived as follows : the set of sequences and/or cycles to be stored is defined as a given set of transitions :

$$\sigma^{k} \rightarrow \sigma^{k+} \quad , \quad k = 1 \text{ to } p.$$

This problem can be stated in a matrix form :

$$C\,\Sigma = \Sigma^{+} ,$$

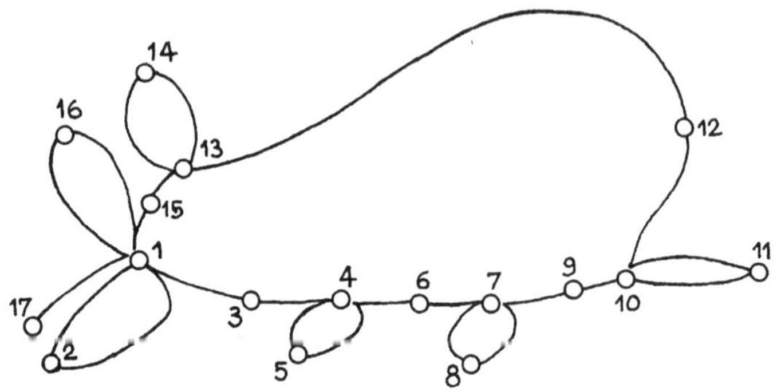

Figure 6 : Sequence storage and retrieval : three sequences are stored, one of which being the poem in the left column ; the network is initialized with the two lines in italics (right column) ; the network retrieves the correct sequences after undergoing five transitions. Bottom : evolution graph, each numeral corresponding to a line.

where C is the (n, n) synaptic matrix, Σ is the (n, p) matrix whose columns are vectors σ^k and Σ^+ is the (n, p) matrix whose columns are vectors σ^{k+}.

For storing sequences without bifurcation points, an adequate solution is given by [4]

$$C = \Sigma^+ \Sigma^I \quad ,$$

where Σ^I is the pseudoinverse of matrix Σ. The storage capacity is O(n).

If bifurcation points are present, the previous expression is no longer satisfactory. A solution is provided by the following transformation : a matrix Γ is defined as the (2n, p) matrix whose columns are the 2n-component vectors defined by the concatenation of vectors σ^k and σ^{k-}, where σ^{k-} denotes the state immediately before σ^k in the sequence to be stored. The (n, 2n) matrix C can then be expressed as

$$C = \Sigma^+ \Gamma^I .$$

The storage capacity of such a network is O(2n), and the retrieval of the stored sequences requires initializing the system with two states.

5. Conclusion

We have presented some problems related to applications of neural networks used as associative memories, and we have shown how these systems can be adapted to specific tasks in pattern recognition and error correction. Clearly, associative memory alone will solve only a part of the very complex problems arising, for instance, in vision. More "biological" types of neural networks, with layered structures, are being investigated, but no firm conceptual foundation for these has been found as yet. However, artificial neural networks have a part to play whenever speed of information retrieval and fault tolerance are of foremost importance. These issues are intimately related to the problem of hardware implementations of neural networks, which has not been discussed in this paper ; this is an important issue if one wants to take full advantage of the capabilities mentioned above. "Neural" silicon chips have already been developed, and many implementations - using the technologies of electronics and of optics - are currently being investigated. The existence of such systems will undoubtedly trigger new advances in this field.

Literature references

1. J.J. Hopfield : Proc. Natl. Acad. Sci. USA 79, 2554 (1982)
2. D. Amit, H. Gutfreund, H. Sompolinsky : Phys. Rev. A 32, 1007 (1985)
   I. Kanter, H. Sompolinsky : Phys. Rev. A 35, 380 (1987)
3. L. Personnaz, I. Guyon, G. Dreyfus : J. Phys. Lett. 46, 359 (1985)
4. L. Personnaz, I. Guyon, G. Dreyfus : Phys. Rev. A 34, 4217 (1986)
5. T. Kohonen : Self Organization and Associative Memories (Springer, Berlin, Heidelberg 1984)
6. L. D. Jackel, R. E. Howard, J.S. Denker, W. Hubbard, S.A. Solla : Neural Networks for Computing (Snowbird, Utah 1987)
7. I. Guyon, L. Personnaz, P. Siarry, G. Dreyfus : In Heidelberg Colloquium on Glassy Dynamics, ed. by J.L. van Hemmen and I. Morgenstern, Lecture Notes in Physics Vol. 275 (Springer, Berlin, Heidelberg 1987)
8. E. Gardner, N. Stroud, D.J. Wallace : Edinburgh preprint 87/394 (1987)
9. L. Personnaz, I. Guyon and G. Dreyfus : Neural Networks for Computing (Snowbird, Utah 1987)

Part V

Physical Devices

Towards the Quantum Computer: Information Processing with Single Electrons

G. Mahler and K. Obermayer

Institut für Theoretische Physik der Universität Stuttgart, Pfaffenwaldring 57, D-7000 Stuttgart 80, Fed. Rep. of Germany

I. Introduction : From Function to Physics

Computational systems can be defined by the task they are expected to perform. At the same time they are physical systems. During the last 50 years, the technology and the architecture of information processing machines led to a drastic miniaturization of the hardware elements by several orders of magnitude. While the length of a vacuum tube in 1940 was about 10cm, the typical length of a transistor on a silicon chip has been reduced to 10μm in the year 1980 and to 1μm in the VLSI-circuits of today. By obvious extrapolation one concludes, that the "nm-chip" containing computing elements of the size of individual molecules will be developed at the beginning of the next century. Does this kind of extrapolation make sense in useful technological terms ? Or does there exist an inherent length scale, below which the constraints imposed on systems capable of information processing must violate the physical laws ?

The physical processes occuring in a molecular computer (i.e. in an information processing system on the nanometer scale) are fundamentally different from processes taking place in a conventional computer. While the dynamical behaviour of a conventional computer is described by collective processes, just a few microscopic degrees of freedom are involved during a computation process in a molecular machine : information processing within a molecular computer may use single electrons, single atoms or single chemical groups. This fact has two remarkable consequences :

1. Molecular information processing systems must be described within the framework of quantum mechanics. Therefore the statistical nature of the measuring process and the coupling to external reservoirs must be taken into account as well as the principal uncertainty and its influence on the preparation process.

2. For decreasing system length it will be more and more difficult to achieve a separation of the microscopic degrees of freedom into the "dynamical" degrees of freedom necessary for the computing process and the "passive" degrees of freedom, which give rise to boundary conditions. We will show that this separation is directly linked to the occurrence of largely different time scales which, in turn, have a structural basis.

II. Multistability and Time-Scales

A switching device can be defined as a bistable system, i.e. a system which is characterized by a repertoire of two distinguishable states. These states must be addressable on a time-scale τ_{relax}, which is small compared to the life-time of these states, τ_d ("dwell-time"). We may thus discriminate between three time-regimes : i. the switching dynamics, $0 \le t \le \tau_{relax}$, ii. the multistable regime, $\tau_{relax} \ll t \ll \tau_d$ and iii. $t > \tau_d$, decay of bistability (thermalization). These features apply to classical switches (compare [1]) and quantum mechanical realizations as well. The latter we model as follows :

We consider an open quantum system I (described by a Hamiltonian $\hat{H}_0$ with its surrounding reservoirs) coupled to an external (macroscopic) preparation system $\hat{H}_p$ and to a (macroscopic) measurement apparatus $\hat{H}_m$. The evolution of I is described by a reduced density matrix $\hat{\rho}$. We will call the stationary states of the system I without any coupling to a preparation and a measurement system <u>attractor states</u> of I. We thus exclude additional stationary attractor states, which may appear in an externally driven system. If the system is coupled to $\hat{H}_p$ ("driving force") only a subset of all possible attractor states remain stationary states of the total system too. Performing a computational step (comp. [2] for classical systems) means driving the system from an initial attractor state $\hat{\rho}_a$ (characterized by expectation values $\{n_a\}$) via certain transient states to a final stationary attractor state $\hat{\rho}_e$ (characterized by expectation values $\{n_e\}$), the result of the computational step. In general not all attractors can be reached from every initial state within one step, but a reliable preparation of the attractor states, which have logical meaning, must be possible using a finite sequence of relaxation processes.

III. Structural Basis of Electronic Multistability

Figure 1a shows the real space model-structure of a single multistable element based on a semiconductor heterostructure. Three blocks of different semiconductor materials A, B and C are embedded into a coating material D. A p-doped layer in the coating material, which is separated from the 3-block structure by an intrinsic spacer region, serves as a source for holes. For an appropriate choice of the semiconductor materials A-D the real space structure leads to a double well structure of the valenceband(VB)- and conductionband(CB)-edges, as shown in Fig. 1b. Holes from the p-doped layer can be trapped within the two VB wells (A and C) if the bandgap-differences between the 3-level block structure and the coating material is large enough. They lead to unoccupied electronic states along the VB-edge (modulation doping).

We describe the model-structure by a simple effective mass envelope function model [3,4] for each band :

$$\left\{-\frac{\hbar^2}{2}\nabla(\underline{m}_n^*)^{-1}\nabla + U_{(\underline{r})}\right\}\psi_{n\underline{k}(\underline{r})} = (E_{n\underline{k}}-E_n)\,\psi_{n\underline{k}(\underline{r})} \tag{1}$$

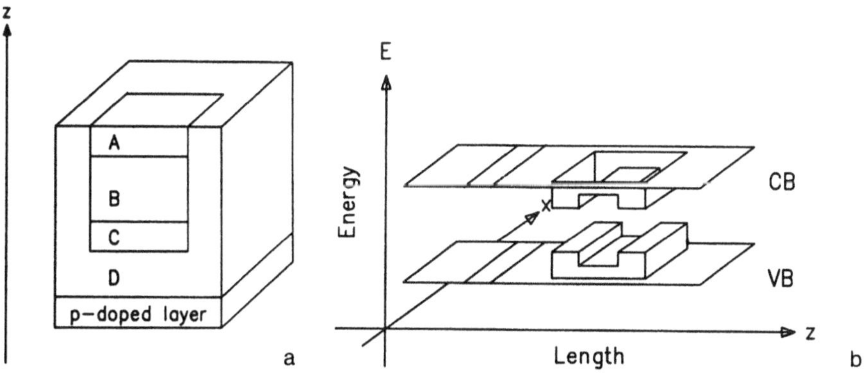

Figure 1 : a) Real space structure of a single multistable element, b) Spatial structure of the VB- and CB-edges within each semiconductor material i. $\psi^{j}_{n\underline{k}(\underline{r})}$ denotes the envelope function for a quasi-particle in the band n with momentum $\underline{k}$, E^{j}_{n} denotes the bulk-bandedge energy within the material j and $U_{(\underline{r})}$ is a slowly varying external potential, e.g. a space charge potential or a Hartree term. The interfaces between the three central blocks are modelled by abrupt steps. To simplify the numerical calculations we describe the interface along the coating material by an infinite barrier, which imposes the boundary condition :

$$\psi_{n\underline{k}}|_{boundary} \equiv 0 \tag{2}$$

on the envelope function.

Figure 2 shows spectrum and eigenfunctions for the lowest lying electron- and heavy hole states of a double quantum box structure realized with the GaAlAs-system. The electronic spectrum consists of discrete electronic energy-levels, characterized by the quantum numbers (k_x, k_y, κ), denoting the momentum of the quasi-particle. In the more realistic case of finite band-discontinuities, the discrete part of the spectrum is embedded in the continuous solid-state spectrum.

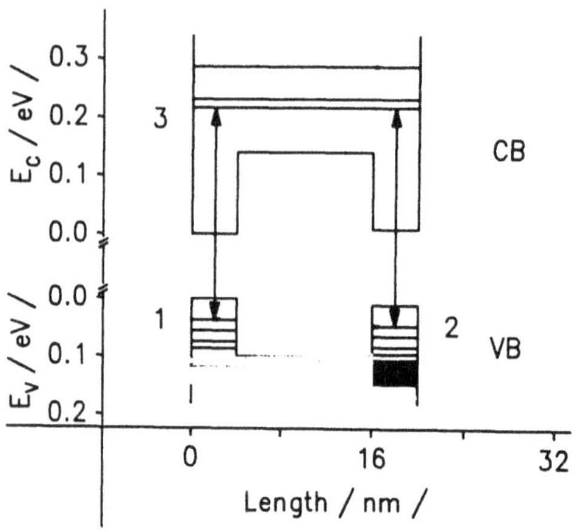

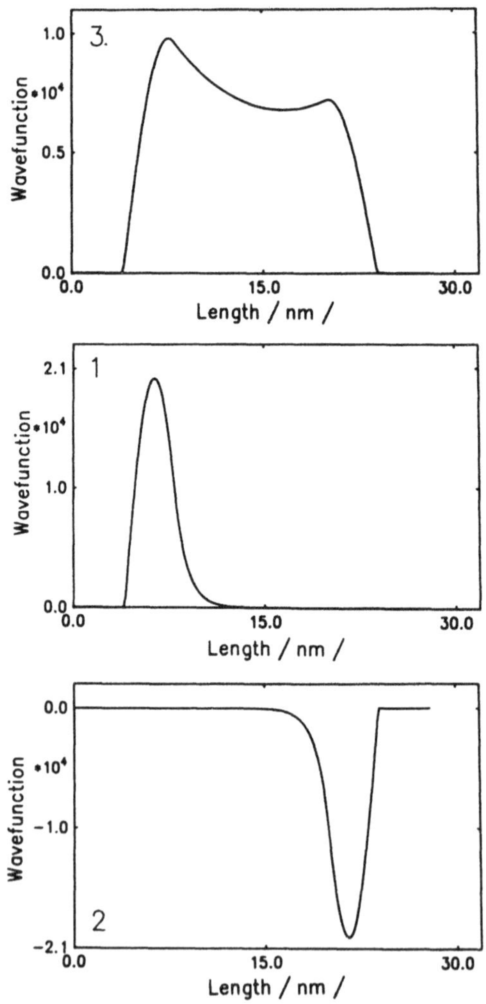

Figure 2: Spectrum and eigenfunctions for a GaAlAs quantum box structure. The geometrical and bandgap parameters of the heterostructure are : d_A = 4nm, d_B = 12nm, d_C = 4nm, S = 10nm, $[Al]_A$ = 0%, $[Al]_B$ = 18%, $[Al]_C$ = 1.8% and bandoffset parameter β = 1.3. Material parameters were taken from [12].

The lowest heavy hole states ("1" and "2") are completely localized within the VB due to their heavy mass, while the lowest electron state ("3") in the CB is completely delocalized over the whole structure. The localization behaviour indicates, that the heterostructure under consideration shows multistationarity. Every distribution of the trapped holes between the two VB-wells should correspond to an attractor state, if the holes within each well are thermalized.

To confirm these assumptions, the coupling of the electronic subsystem to the other degrees of freedom of the heterostructure and its surrounding must be considered. Neglecting correlations, this can be done by describing the surroundings as external reservoirs. Coupling to reservoirs leads to transitions between the electronic eigenstates of the heterostructure : band-band transitions, inter-well transitions (i.e. transitions between eigenstates with different localization behaviour within the same band) and intra-well transitions.

The band-band transitions are responsible for the computational dynamics. For direct semiconductors with a large bandgap the corresponding time-scale is determined by optical decay. The inter-well processes are responsible for the stability of the attractor states. The corresponding time-scale is determined by electron-phonon scattering processes and can be estimated by considering the first order intraband deformation potential scattering by acoustic (bulk-) phonons for the heavy hole states. This will give the appropriate time-scale for low temperature and acoustically well matched materials, when optical phonon scattering and scattering by slab- and interface modes can be neglected. An upper bound for these transition probabilities may be shown to be given by [5] :

$$
\begin{aligned}
w^{opt}_{n\underline{k},n'\underline{k}'} &\propto |\int \chi^*_{n\kappa}\chi_{n'\kappa'}dz|^2 \\
w^{ph}_{n\underline{k},n\underline{k}'} &\propto \{\int |\chi^*_{n\kappa}||\chi_{n\kappa'}|dz\}^2.
\end{aligned}
\tag{3}
$$

Since the transition rates are predominantly determined by the spatial overlap of the envelope functions it may be possible to tailor desired selection rules for transitions between the electronic eigenstates by an appropriate choice of the geometrical and bandgap-parameters.

Fig. 3 shows the overlap-integrals in equation (3) depending on selected geometrical parameters of a model heterostructure. With increasing complexity of the real-space structure, we get an increasing separation of the overlap-integrals for the different transitions. For an appropriate choice of the system-parameter one may achieve a difference of several orders of magnitude between the overlap-integrals 3-1 and 3-2 and the overlap-integral 2-1, respectively. This leads to a large separation of time-scales between the time-scale of the computational dynamics, τ_{relax}, and the time-scale of stability, τ_d as shown in Table 1 for the GaAlAs-system. The inter-well scattering rate between electronic eigenstates deeper in the VB is of course much greater than the scattering rate 2 → 1. If the double quantum box has trapped just one hole, their contribution to the total

Table 1 : Transition probabilities for the electronic transitions 3-1, 3-2 and 2-1 for a GaAlAs quantum box structure. The geometrical and bandgap-parameter are $d_A = d_C = 4$nm, $d_B = 12$nm, $[Al]_A = 0\%$, $[Al]_B = 20\%$, $[Al]_C = 1\%$, T = 10K and S = 10nm. Material parameter were taken from [12].

β	$3\text{-}1(\tau^{-1}_{relax})$	$3\text{-}2(\tau^{-1}_{relax})$	$2\text{-}1(\tau^{-1}_{d})$
1.0	4×10^7/s	3×10^7/s	4×10^{-2}/s
1.5	7×10^7/s	3×10^7/s	4/s
2.0	1×10^8/s	3×10^7/s	130/s

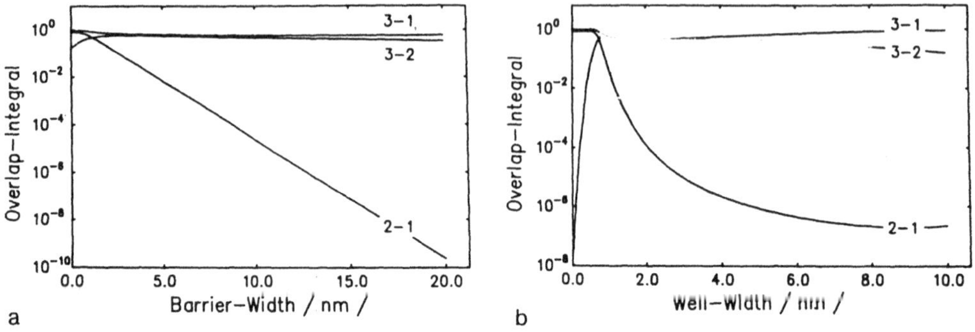

Figure 3: Overlap-integrals as a function of barrier-width and well-width for two semiconductor model structures. a) Overlap-integrals 3-1, 3-2 und 2-1 as a function of the barrier-width d_B for $d_A = d_C = 4$nm, $V_{VB}^A = 0$meV, $V_{VB}^B = 100$meV, $V_{VB}^C = 10$meV, $V_{VB}^D = 200$meV and bandoffsetparameter $\beta = 0.5$. The effective mass was $0.7m_0$ for the heavy holes in the VB and $0.07m_0$ for the electrons in the CB. b) Overlap-integrals 3-1, 3-2 und 2-1 as a function of the well-width $d_A = d_C$ for $d_B = 12$nm, $V_{VB}^A = 0$meV, $V_{VB}^B = 100$meV, $V_{VB}^C = 10$meV, $V_{VB}^D = 200$meV and bandoffset-parameter $\beta = 0.5$. The effective mass was the same as in (a).

scattering rate between two attractor states decreases exponentially for low temperatures. In the limit T = 0K it will be zero, since all excited (heavy hole and light hole) states are unoccupied.

In the pertinent model structure multistationarity is achieved by constructing systems showing a time-scale hierarchy. On the time-scale of optical decay the system may follow only selected pathways in state-space (e.g. the direct transition 2 → 1 is forbidden, electron transfer is possible only via level-3). This is accomplished by exploiting the hierarchical nature of various degrees of freedom, since a separation into "dynamical" and "passive" degrees of freedom is still possible for quantum boxes. A heterostructure as described in Fig. 2 still contains approximately 10^4 - 10^5 atoms and 10^5 - 10^6 electrons, most of them are used to build up the structure. Their influence on the "switching electrons" is described within the framework of the effective mass theory by an effective "potential", which is not severely disturbed during the switching dynamics of a few electrons, and by electronic transitions due to electron-electron and electron-phonon interaction. In the case of electron-phonon scattering it was shown, that "dynamical" and "passive" degrees of freedom decouple on the time-scale of computation for an appropriate choice of the system's parameter. Electron-electron interactions have not been discussed; they must be taken into account within a refined theory.

IV. Switching Dynamics : Effective One-Particle Model

1. Equation of Motion for the Reduced Density Matrix

In the following we will consider a heterostructure, which has trapped just a single hole, i.e. a heterostructure where the Fermi-energy level was pinned between level-1 and level-2 (Fig. 2). Note that the energy levels 1, 2 and 3 are not degenerate. Spin degeneracy is lifted by spin-orbit coupling and by the spatial confinement of the quasi-particles [6]. Since for low temperatures the hole will occupy only the highest VB state in each quantum box, we can describe the dynamical behaviour and the coupling to the preparation system with a simple 3-level model, containing the highest VB-states 1 and 2 and the lowest CB-state 3. The preparation system induces transitions between the energy levels 2, 1 and 3 via single modes of the electromagnetic field. Since we are not explicitly concerned with the dynamics of the preparation system we describe the electronic subsystem during preparation as an externally driven system with a time-dependent coupling Hamiltonian $\hat{H}_{0-p}$:

$$\hat{H}_{0-p} = 2\hbar \sum_{j=1}^{2} \{\underline{A}_j \cos(\omega_j t+\phi_j)\} \sum_{m=1}^{2} \{\underline{G}_m \hat{a}_m^+ \hat{a}_3 + \underline{G}_m^* \hat{a}_3^+ \hat{a}_m\} \tag{4}$$

where the state of the light field is described by up to two phase-correlated classical plane waves. The spontaneous emission modes constitute a totally absorbing external reservoir, whose effects can be taken into account (via the Markoff-approximation) by relaxation terms in a generalized master equation [7] :

$$\frac{d\hat{\rho}}{dt} = -\frac{i}{\hbar} [\hat{H}_0+\hat{H}_{0-p},\hat{\rho}] + \sum_{j=1}^{2} \frac{w_j}{2} \{[\hat{a}_j^+\hat{a}_3,\hat{\rho}\hat{a}_3^+\hat{a}_j] + [\hat{a}_j^+\hat{a}_3\hat{\rho},\hat{a}_3^+\hat{a}_j]\} . \tag{5}$$

$\hat{H}_0$ is the Hamiltonian of the electronic subsystem. In general a single particle theory is not sufficient and electron-electron coupling must be included, but in a first approximation we may assume the effect of the electron-electron interaction to be included in a renormalization of the single particle eigenenergy. This will be a good approximation in the case of a single trapped hole, if the transitions between the electronic eigenstates are off-resonance. Therefore we get :

$$\hat{H}_0 = \sum_{j=1}^{3} E_j \hat{a}_j^+ \hat{a}_j \tag{6}$$

where E_j is the renormalized energy of the electronic eigenstates. Since we are interested in the dynamical behaviour of the system on the time-scale of the computational processes, we need not consider the relaxation processes 2→1 by phonon scattering. Using the transformation :

$$\bar{\Omega}_i = \frac{\underline{G}_i \underline{e}_i}{\omega_i} e^{-i\theta_i}$$

$$\rho_{13} = \bar{\rho}_{13} e^{i(\omega_1 t+\phi_1+\theta_1)}$$

$$\rho_{23} = \bar{\rho}_{23} e^{i(\omega_2 t+\phi_2+\theta_2)}$$

$$\rho_{12} = \bar{\rho}_{12} e^{i((\omega_1-\omega_2)t+(\phi_1-\phi_2)+(\theta_1-\theta_2))} \tag{7}$$

we get within a generalized rotating wave approximation for the transformed elements of the density matrix (in the basis of the electronic eigenstates of $\hat{H}_0$) :

$$
\begin{vmatrix} \dot{\rho}_{11} \\ \dot{\rho}_{22} \\ \dot{\rho}_{33} \\ \dot{\bar{I}}_{13} \\ \dot{\bar{R}}_{13} \\ \dot{\bar{I}}_{23} \\ \dot{\bar{R}}_{23} \\ \dot{\bar{I}}_{12} \\ \dot{\bar{R}}_{12} \end{vmatrix}
=
\begin{vmatrix}
0 & 0 & w_1 & -2\bar{\Omega}_1 & 0 & 0 & 0 & 0 & 0 \\
0 & 0 & w_2 & 0 & 0 & -2\bar{\Omega}_2 & 0 & 0 & 0 \\
0 & 0 & -(w_1+w_2) & 2\bar{\Omega}_1 & 0 & 2\bar{\Omega}_2 & 0 & 0 & 0 \\
\bar{\Omega}_1 & 0 & -\bar{\Omega}_1 & -\frac{1}{2}(w_1+w_2) & \delta_1 & 0 & 0 & 0 & \bar{\Omega}_2 \\
0 & 0 & 0 & -\delta_1 & -\frac{1}{2}(w_1+w_2) & 0 & 0 & -\bar{\Omega}_2 & 0 \\
0 & \bar{\Omega}_2 & -\bar{\Omega}_2 & 0 & 0 & -\frac{1}{2}(w_1+w_2) & \delta_2 & 0 & \bar{\Omega}_1 \\
0 & 0 & 0 & 0 & 0 & -\delta_2 & -\frac{1}{2}(w_1+w_2) & \bar{\Omega}_1 & 0 \\
0 & 0 & 0 & 0 & \bar{\Omega}_2 & 0 & -\bar{\Omega}_1 & 0 & \delta_1-\delta_2 \\
0 & 0 & 0 & -\bar{\Omega}_2 & 0 & -\bar{\Omega}_1 & 0 & -(\delta_1-\delta_2) & 0
\end{vmatrix}
\cdot
\begin{vmatrix} \rho_{11} \\ \rho_{22} \\ \rho_{33} \\ \bar{I}_{13} \\ \bar{R}_{13} \\ \bar{I}_{23} \\ \bar{R}_{23} \\ \bar{I}_{12} \\ \bar{R}_{12} \end{vmatrix}
\qquad (8)
$$

where ρ_{ii} denote the diagonal elements of the density matrix and $\bar{R}_{ij}$ and $\bar{I}_{ij}$ the real and imaginary part of the transformed off-diagonal elements respectively.

2. Attractor states of the 3-level system

The attractor states are by definition the stationary states of the electronic subsystem without coupling to the preparation and the measurement system. We must therefore set $\bar{\Omega}_1 = \bar{\Omega}_2 = 0$ and $\delta_1 = \delta_2 = 0$ within (8). The resulting equations may readily be integrated and we get for $t \to \infty$: ρ_{11}, ρ_{22}, $\bar{R}_{12}$ and $\bar{I}_{12} \neq 0$ depending on the initial conditions, and all other matrix elements equal to zero. Since there is no relaxation process $2 \to 1$, a coherent superposition of the eigenstates |1> and |2> does not decay. The attractor states can be classified into three groups :

1. eigenstates of $\hat{H}_0$ with $\bar{\rho}_{ij} = \delta_{i1}\delta_{j1}$ or $\bar{\rho}_{ij} = \delta_{i2}\delta_{j2}$ ("deterministic" attractors).

2. a coherent superposition of the eigenstates |1> and |2> with $\rho_{11} \neq \rho_{22} \neq 0$ and $\bar{\rho}_{12} \neq 0$, where the density matrix after diagonalization contains only one element, which is not zero ("coherent" attractors).

3. a stationary state with $\rho_{11} \neq \rho_{22} \neq 0$ and $\bar{\rho}_{12} \neq 0$, where the density matrix after diagonalization contains more than one element, which is not zero ("stochastic" attractors).

Note that these attractor states are quite different from the stationary states of conventional information processing systems, since they are metastable states of an isolated system, which is not coupled to an external driving apparatus. Therefore no dissipation of energy is needed to maintain its stationary state.

3. Subset of Attractor States : Alphabet

We first discuss a 3-level system, which is driven by a single light mode. The light mode induces transitions between the eigenstates |1> and |3> (Fig. 4 inset), and the dynamic of the 3-level system is described by (8) with $\bar{\Omega}_2 = 0$ and $\delta_2 = 0$. In this case the only stationary state is given by $\rho_{ij} = \delta_{i2}\delta_{j2}$ independent of the choice of the parameters, i.e. all off-diagonal elements decay. Therefore 1-mode lightpulses may be used to reliably prepare "deterministic" attractor states, since the preparation process is insensitive to "errors" in the coupling hamiltonian $\hat{H}_{0-p}$. 1-mode lightpulses provide a reliable error correction mechanism too, since the final state does not depend on the initial state of the system. Fig. 4 shows the diagonal elements ρ_{11}, ρ_{22} and ρ_{33} as a function of time during a switching process. We may define a switching time T by :

$$T = \mathrm{Max}\left\{-(\mathrm{Re}\lambda_j)^{-1}\right\} \tag{9}$$

where λ_j are the (complex) eigenvalues of the matrix in equation (8). For high values of $\bar{\Omega}_1/(w_1+w_2)$ the switching time is limited by the spontaneous emission $3 \rightarrow 2$ and attains a minimal value $T = 2/w_2 \approx \tau_{relax}$ independent of the detuning. The system remains in this attractor state even after the light pulse is turned off.

Let, finally, the state of the preparation system consist of two different light modes, which induce transitions between the eigenstates |1>, |3> and |2>, |3> respectively (Fig. 5b inset). The system must now be described by the whole set of (8). If the two light modes are "symmetrically" detuned, i.e. if $\delta_1 = \delta_2$, the stationary state is given by :

$$\rho_{11} = \frac{\bar{\Omega}_2^2}{\bar{\Omega}_1^2+\bar{\Omega}_2^2} \qquad \rho_{22} = \frac{\bar{\Omega}_1^2}{\bar{\Omega}_1^2+\bar{\Omega}_2^2} \qquad R_{12} = \frac{\bar{\Omega}_1\bar{\Omega}_2}{\bar{\Omega}_1^2+\bar{\Omega}_2^2} \; . \tag{10}$$

All other matrix elements are zero. The resulting stationary state is a coherent attractor state of the 3-level system. Fig. 5 shows the time-dependence of the diagonal and some of the off-diagonal elements of the density matrix during a preparation process. The population of level-3 is zero in the stationary state, although light is still irradiated. This effect is known as coherent population trapping and is found only if the driving light modes are correlated [8]. If the two light modes are asymmetrically detuned, the population of level-3 is no longer zero and the stationary state is no longer an attractor state of the isolated system. A complete switching process in this case consists of a light pulse of

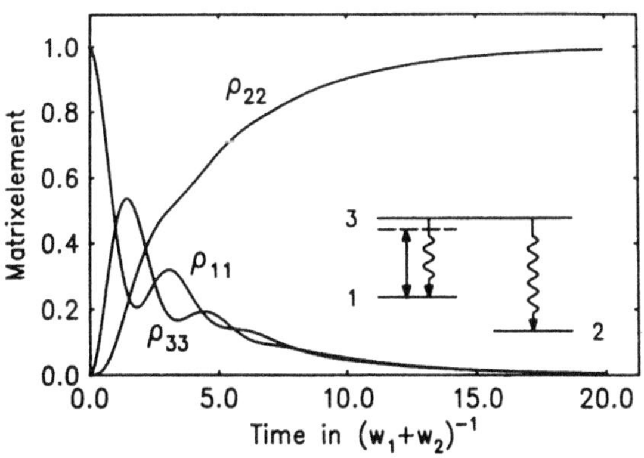

Figure 4: Diagonal elements of the density matrix as a function of time for a dissipative 1-mode process for $\bar{\Omega}_1 = w_1+w_2$, $\delta_1 = 0$, $w_1 = w_2$, initial condition : $\rho_{ij} = \delta_{i1}\delta_{j1}$. The inset shows the quantum optical model.

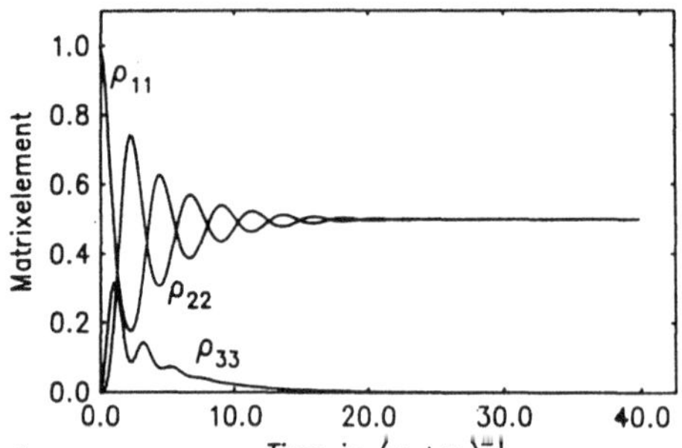

a

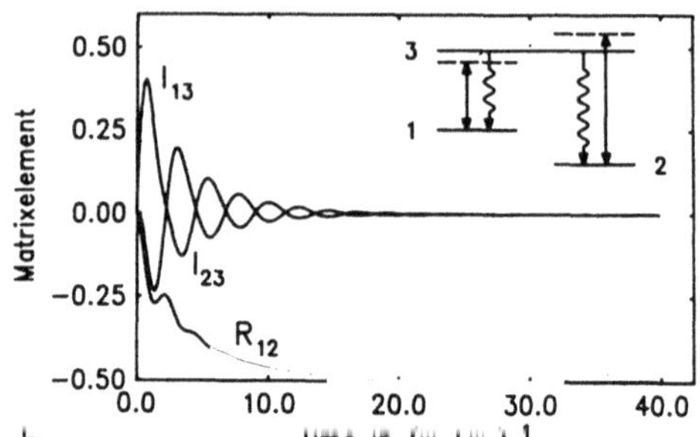

b

Figure 5 : Diagonal (a) and off-diagonal (b) elements of the density matrix as a function of time for a dissipative 2-mode process for $\bar{\Omega}_1 = \bar{\Omega}_2 = w_1+w_2$, $\delta_1 = \delta_2 = 0$, $w_1 = w_2$, initial condition $\rho_{ij(0)} = \delta_{i1}\delta_{j1}$. The inset in (b) shows the quantum optical model.

finite duration followed by a relaxation process (i.e. $\bar{\Omega}_1 = \bar{\Omega}_2 = 0$), which drives the system into a stochastic attractor state. Since the resulting attractor state depends on the light pulse parameters, the preparation process is more sensitive against errors than in the case of preparation by 1-mode pulses. The final state is still independent of the initial conditions, so that dissipation provides an reliable error correction mechanism too.

4. Measurement of Deterministic Attractor States

Since the whole quantum box structure containing trapped holes is no longer electrically neutral, it carries an electric dipol moment. During a switching process charge is transferred from one VB-well into the other, and the electric dipole moment is changed. In an experimental situation measurement must be performed on an ensemble of multistable elements.

Single multistable elements may be detected via optical measurements. Therefore the heterostructure must have a relevant subspectrum as shown in Fig. 6. The localization behaviour of the enevelope function is indicated. Measurement is performed by irradiation of light inducing the transitions 1-4 and 2-5 and detecting the spontaneous emitted photons of the corresponding energy. The sensitivity of measurement is limited by the time-scale of stability of the attractors. For attractors with long enough life times single electrons may be detected [9].

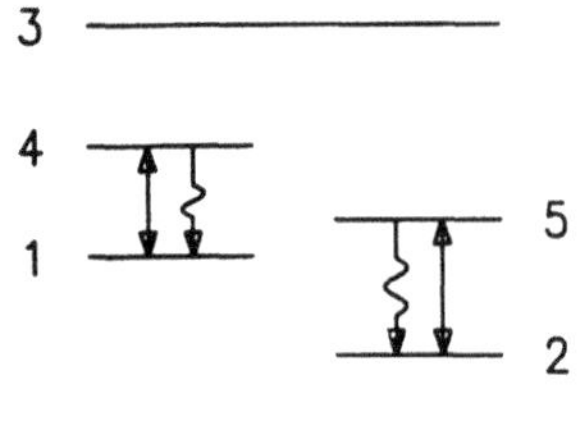

Figure 6: Relevant subspectrum of a heterostructure, whose attractor states can be detected via optical measurement. The length of the bars indicates the localization behaviour of the corresponding envelope functions.

V. Switching Dynamics : Effective Few-Particle Model

For the realization of logical elements or other information processing units it is necessary, that the "switching" electrons within a multistationary system interact. There are four types of interactions possible : coupling via the Pauli-principle, the Coulomb-interaction and coupling via the modes of the electromag-

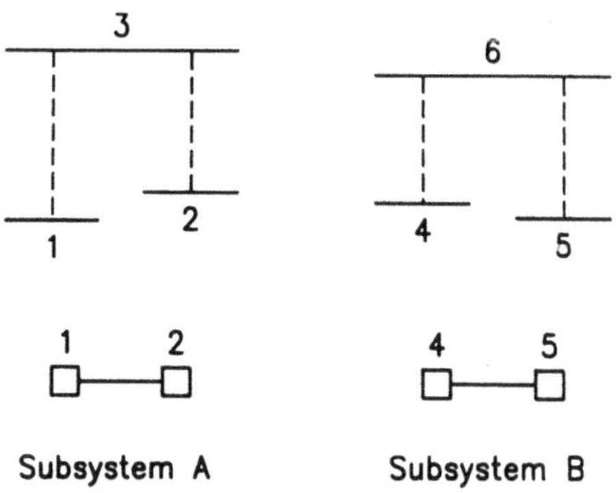

Figure 7: Quantum optical model and coupling scheme for an information transfer element.

netic field or the phonon-system. The latter types of interaction are weak and may be used for information processing only for large ensembles of multistable elements. Interaction via the Pauli-principle and the Coulomb-interaction, on the other hand, may provide a reliable computational mechanism based on individual electrons. Let us briefly discuss the coupling via Coulomb-interaction, i.e. a non-local operator.

Let us imagine two bistable subsystems A and B which are separated by a distance large enough so that the overlap between the envelope functions of subsystems A and B can be neglected (Fig. 7). Each subsystem is occupied by one electron. We assume, that the potential barrier between the two subsystems is high enough, so that an electron (on the time-scale of computation) can not move from one subsystem to the other.

The Coulomb matrix element reads :

$$W(ijkl) = [\frac{e^2}{4\pi\varepsilon\varepsilon_0}] \iint d^3x d^3x' \; \Psi_i^*(\underline{x})\Psi_j^*(\underline{x}')\frac{1}{|\underline{x}-\underline{x}'|}\Psi_k(\underline{x}')\Psi_l(\underline{x}) \tag{11}$$

where Ψ_i is the electronic wavefunction of state i and ε the dielectric constant of the embedding semiconductor. For j=k and i=l the Coulomb matrixelement (11) describes the renormalization of the energy levels of system A depending on the actual state of system B and vice versa. If we choose $\underline{R}$ to be the distance $\underline{R}_{ij}$ between the centres of charge $\underline{R}_i$ and $\underline{R}_j$ the renormalization energy W_{ijij} may be approximated by [10] :

$$W(ijij) \approx \frac{e^2}{4\pi\varepsilon\varepsilon_0 R} \; . \tag{12}$$

For j≠k and i≠l real and virtual electronic transition processes occur, which will be considered elsewhere [10].

Since the Coulomb renormalization depends on the distance R_{ij} between the centres R_i and R_j of the charge distributions, the transition frequencies ω_{64} and ω_{65} of subsystem B are different for the two possible switching states of A. For a distance of ca. 50nm between two double quantum box systems as described in Fig. 4 the difference in the transition energies is about 0.1meV - 1meV. These energy differences are still large enough compared with the linewidth of a typical laser pulse, so that each transition (A=1 and A=2) may be addressed separately.

Fig. 7 shows the quantum optical model and the coupling scheme of an information transfer element. Four out of six possible (deterministic) attractor states have logical meaning :

$$
\begin{array}{llll}
a_1 = + + - - & & f_{(a_1)} = \text{not defined} & \\
a_2 = + - + - & & f_{(a_2)} = 0\ 1 & \\
a_3 = + - - + & & f_{(a_3)} = 0\ 0 & \\
a_4 = - + + - & & f_{(a_4)} = 1\ 1 & \\
a_5 = - + - + & & f_{(a_5)} = 1\ 0 & \\
a_6 = - - + + & & f_{(a_6)} = \text{not defined}\ . & \qquad (13)
\end{array}
$$

Let us assume, that the data element A contains the information to be copied into the data element B. A complete transfer operation now consists of two (dissipative) light pulses with the transition frequencies $\omega_{65}(A{=}1)$ and $\omega_{64}(A{=}2)$ (Fig. 8) :

$$
\begin{array}{lcl}
 & \omega_{65},\ \omega_{64} & \\
a_2 & \text{———} & a_2 \\
a_3 & \text{———} & a_2 \\
a_4 & \text{———} & a_5 \\
a_5 & \text{———} & a_5
\end{array}
\qquad (14)
$$

which leads to the logical table :

initial state		final state	
a_3	0 0	0 0	a_2
a_2	0 1	0 0	a_2
a_5	1 0	1 1	a_5
a_4	1 1	1 1	a_5 .

(15)

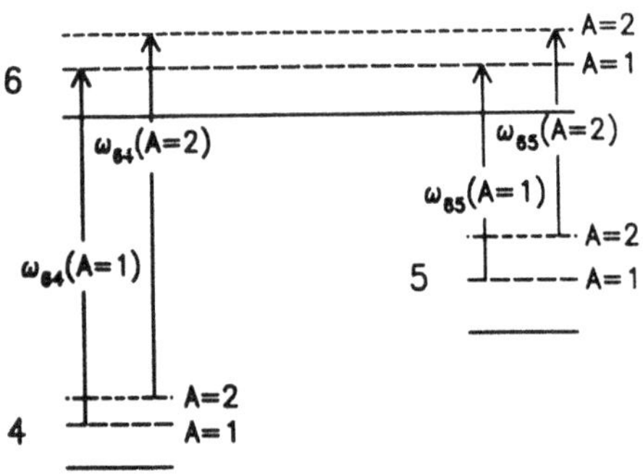

Figure 8: Information transfer between two 3-level systems.

VI. Architecture

In principle, the Coulomb renormalization may be used to perform other logical operations too, as was proposed by [11] in the case of switching molecules. This leads to a sequential computational dynamics based on localized "molecules". A different architecture able to reduce the communications with the macroscopic environment can be characterized as "distributed" computation. This is based on a cellular structure, in which quasi-molecular multistable elements are repeated many times. Such a configuration comes close to Hopfield-type models extensively discussed in this Symposium. The preference of a certain architecture may therefore not only result from the implied task ("function") but also from the underlying physical properties of the computational system. It is remarkable that for such quasi-molecular cellular automata the external light-pulses do not only serve as a "clock" but at the same time define the "transition rules". As all cells are updated simultaneously and within a repetition rate large compared with the reciprocal dwell-time, τ_d, the finiteness of τ_d does no longer play a significant role at least during the computational process [10].

Preliminary results of this investigation have been published in [13] and reviewed in [14]. The general issue of the physical background of computation has been brought to our attention mainly by the fundamental work of R. Landauer [15],[16]. Financial support by the Volkswagenstiftung and the Deutsche Forschungsgemeinschaft is gratefully acknowledged.

References :

1 Ebeling U. and Feistel R. : Physik der Selbstorganisation und Evolution, Akademie Verlag Berlin 1982, page 303

2 Haken H. : in Stochastic Phenomena and Chaotic Behavior in Complex Systems ed. by P. Schuster, Springer 1984

3 Altarelli M. : in Heterojunctions and Semiconductor Superlattices (Ed. Allen G., Bastard G., Boccara N., Lannoo M. und Voos M.), Springer-Verlag Berlin (1986)

4 Asada M., Miyamoto Y. and Suematsu Y., IEEE J. Quant. El. QE-22(9), 1915 (1986)

5 Obermayer K., Teich W. and Mahler G., to be published

6 Lassnig R., Phys. Rev. B31(12), 8076 (1985)

7 Haken H. : Quantenfeldtheorie des Festkörpers, B. G. Teubner-Verlag Stuttgart, p. 293ff (1973)

8 Dalton B. J. and Knight P. L., J. Phys. B15, 3997 (1982)

9 Nagourney W., Sandberg J. and Dehmelt H., Phys. Rev. Lett. 56(26), 2797 (1986)

10 Teich W., Obermayer K. and Mahler G., to be published

11 Hadden R. C. and Stillinger F. H., in Molecular Electronic Devices (Ed. Carter L.), Marcel Dehler Inc. (1982)

12 Adachi S., J. Appl. Phys. 58(3), R1 (1985)

13 Obermayer K., Mahler G. and Haken H., Phys. Rev. Lett. 58, 1792 (1987)

14 Maddox J., Nature 327, 97 (May 1987)

15 Landauer R., Foundations of Physics 16, 551 (1986)

16 Landauer R. and Buettiker M., Physica Scripta T9, 155 (1985)

Chaos and Pulsating Instabilities in Lasers

R.G. Harrison, J.V. Moloney, and J.S. Uppal

Department of Physics, Heriot-Watt University,
Edinburgh EH14 4AS, United Kingdom

1. Introduction

In recent years radically new mathematical insights into the dynamical behaviour of many nonlinear systems show that along with yielding regular and repeatable behaviour, many nonlinear systems also exhibit unstable, even chaotic, solutions. Furthermore, the transition from stable to chaotic behaviour, which may occur when varying a control parameter of the system, follows specific, well-defined routes which are universal in the sense that they are independent of the physical properties of the system they describe. It is these signatures which have been a major impetus to experimentalists in the subsequent search for physical systems that exhibit these phenomena. Such phenomena exist in optics; both lasers (active systems), in which the optical signal is derived from stimulated emission generated within an optical cavity containing a gain medium, and passive systems for which the optical signal is but the transmission of an input light signal through an optical cavity containing a nonlinear medium. The latter are being increasingly recognised for their potential application as bistable all-optical logic elements.

Of the variety of physical systems that exhibit deterministic instability phenomena, optical systems, both lasers and passive devices, provide attractive systems for quantitative investigation due to their simplicity both in construction and in the mathematics that describe them.

2. Universality in Chaos

The discovery by Lorentz [1] in 1963 that even a simple set of three coupled first-order, nonlinear differential equations can lead to completely chaotic trajectories is recognised as a landmark and is fundamental to our understanding of laser instabilities.

Unlike linear systems, nonlinear systems must be treated in their full complexity, and so there is no general analytical approach for solving them.

The temporal evolution in the behaviour of such systems can be characterised when presented as a trajectory of a point in the phase space of its dynamical variables. If an initial condition of a dissipative nonlinear dynamical system such as a laser is allowed to evolve for a long time, the system, after all the transients have died out, will eventually approach a restricted region of the phase space called an attractor. A dynamical system can have more than one attractor, in which case different initial conditions lead to different types of long-time behaviour.

The simplest attractor in phase space is a fixed point; the nonlinear system is attracted towards this point and stays there. For other control conditions the system may end up making a periodic motion. The limit or attractor of this motion is a periodic cycle called a limit cycle. However, when the operating conditions exceed a certain critical value, the periodic motion of the system breaks down into a more complex chaotic pattern which never repeats itself. This motion represents a third kind of attractor in phase space called a chaotic or strange attractor. These various types of attractor are shown in Fig. 1.

A trajectory on a chaotic attractor exhibits most of the properties intuitively associated with random functions, although no randomness is ever explicitly added. The equations of motion are purely deterministic; the random behaviour emerges spontaneously from the nonlinear system. Over short times, the trajectory of each point can be followed, but over longer periods small differences in position are greatly amplified making the predictions of long-term behaviour impossible. As such, arbitrarily close initial conditions can lead to trajectories which after a sufficiently long time diverge widely, in marked contrast to that of the fixed point and limit cycle attractors which settle down to the same solutions.

Erratic and aperiodic temporal behaviour of any of the systems' variables implies a corresponding continuous spectrum for its Fourier transform which is, therefore, also a further signature of chaotic motion. However, other factors, including noise, can lead to continuous spectra, and distinguishing chaos from noise is one of the major problems of the field. Hence, although time series,

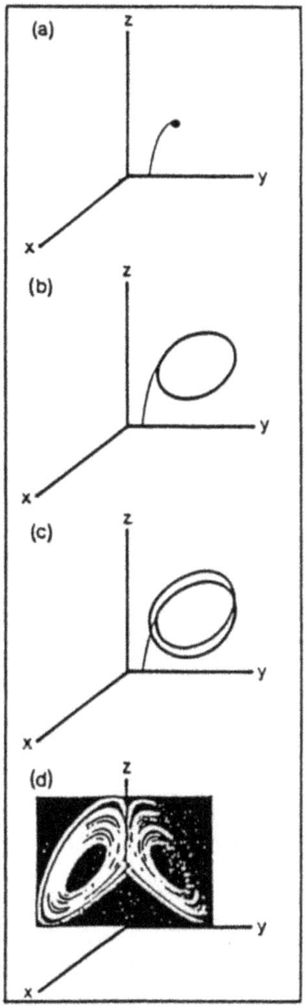

Fig. 1. Phase space portraits of the attractor for the dynamical variables x, y and z. a) Stable point corresponding to a steady state in time after initial transients have died out shown as the thin continuous line. b) Period-one limit cycle corresponding to a periodic solution in time of single frequency. c) Bifurcation to period-two limit cycle; a periodic solution in time with double the period of that in b). Successive period doubling bifurcations lead to an eventual chaotic solution. d) Shows the strange or chaotic attractor for a Lorenz-Haken system describing a single mode laser with a homogeneously broadened two-level gain medium.

power spectra and routes to chaos (see below) collectively provide strong evidence of deterministic behaviour, further signatures are desirable for its full characterisation and in discriminating it from stochastic behaviour. Here analysis of trajectories of a point in the phase space of its dynamical variables is required. However, for a system with, say, N degrees of freedom it seemed that it would be necessary to measure N independent variables; an awesome if not impossible task for complex system. Consequently mathematicians have long tried to develop practical techniques for extracting specific finite dimensional information from the limited output provided by experiment, typically the time record of a specific physical observable; that is, one variable of the system, e.g. laser intensity. Here embedding theorems [2] have been recently used to reconstruct phase portaits from which Lyapunov exponents may be determined that measure the average rate of exponential separation or contraction of nearby points on the attractor. These measure intrinsically dynamical properties, unlike power spectra, and provide quantitative measures by which chaotic motion may be distinguished from stochastic behaviour.

The discoveries that deterministic chaos proceeds through a limited number of specific routes is profoundly significant as such behaviour is not restricted to a particular model description of a particular physical system. Rather, nonlinear physical systems in all branches of science which may be formally described by the same set of mathematical equations will give solutions that evolve identically in time through one or other routes to chaotic motion. The unique effect of such unification between many separate scientific disciplines forms the basis for the foundation of synergetics [3,4]. There are at least three common routes referred to as period doubling, intermittency and two-frequency scenarios.

Period doubling. From considering various difference equations, many of which can be reduced to simple one-dimensional maps, solutions have been found to oscillate between stable values, the period of which successively doubles at distinct values of the external control parameter [5]. This continues until the number of fixed points becomes infinite at a finite parameter value, where the variation in time of the solutions becomes irregular. One example showing such behaviour is the simple logistic map

$$X_{n+1} = r X_n (1 - X_n) .$$

Many complex physical systems, often described by large numbers of coupled differential equations, may be reduced in some conditions to this or similar maps. Period doubling bifurcation has been experimentally observed in numerous systems.

Intermittency. Intermittency [6] means that a signal which behaves regularly in time becomes interrupted by statistically-distributed periods of irregular motion. The average number of these intermittent bursts increases with the external control parameter until the condition becomes completely chaotic.

Two frequency. Turbulence in time was originally considered as a limit of an infinite sequence of instabilities (Hopf bifurcations) evolving from an initial stable solution each of which creates a new basic frequency. However, it has been recently shown [7] that after only two or perhaps three instabilities in the third step the trajectory becomes attracted to a bounded region of phase space in which initially closed trajectories separates exponentially; as such the motion becomes chaotic.

3. Chaos in Lasers

Two-level laser

Chaotic behaviour in lasers may exist in even the simplest of systems; one in which population inversion is established between two discrete energy levels of the medium and where the lasing transition between these two levels is homogeneously broadened. A further simplification is that the laser cavity, a Fabry-Perot, or ring resonator system surrounding the gain medium, be sufficiently short so that only one resonant frequency of the cavity lies within the bandwidth of the gain medium and that this mode be resonantly tuned to the gain centre frequency. The frequency spacing (Δv) between cavity modes for a Fabry-Perot cavity is given by

$$\Delta v = cn/2L$$

where c is velocity of light, n the refractive index of the lasing medium and L the cavity length (see Fig. 2). For conditions in which the gain or population inversion is maintained at a constant level by, for example, constant electrical or optical excitation, and for the single mode system, lasing occurs with a constant output power at the frequency of the single cavity mode.

However, the discovery that for certain operating conditions emission could be periodic or even chaotic implies that the signal comprises more than one frequency, contrary to the accepted understanding of single-mode operation. Predictions of such behaviour were initially identified by Haken [8] through the mathematical equivalence of the equations describing laser action, the Maxwell-Bloch equations, and those derived earlier by Lorenz to describe chaotic motion in fluids.

General prerequisites for the onset of deterministic chaos include: that apart from a nonlinear interaction there is a sufficiently large phase space, the minimum requirement being that the system possesses at least three degrees of freedom.

The Maxwell-Bloch equations described below for the special case of a single-mode laser with field tuned to the centre of the gain line such that both field and polarisation are real quantities, satisfy the minimum condition of three independent variable equations, each of which has its own relaxation.

$$\frac{dE}{dt} = -\kappa E + \kappa P$$

$$\frac{dP}{dt} = \gamma_\perp ED - \gamma_\perp P$$

$$\frac{dD}{dt} = \gamma_{||}(\lambda + 1) - \gamma_{||}D - \gamma_{||}\lambda EP$$

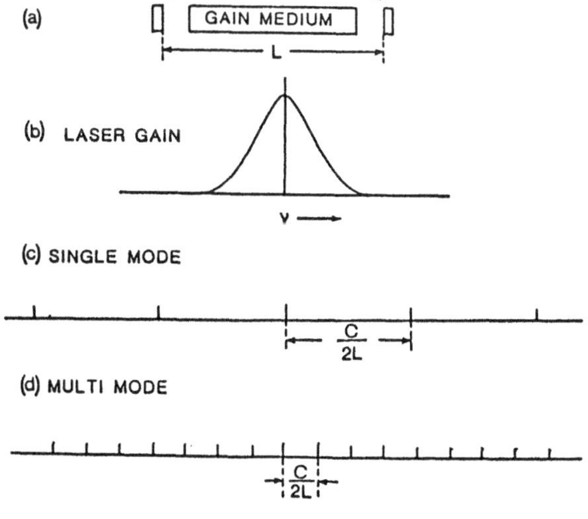

Fig. 2. a) A Fabry-Perot laser cavity system with a partially transmitting mirror for coupling out the laser emission; b) Lorentzian gain profile for a simple two-level homogeneously broadened lasing medium for which each and every atom/molecule emits identically. c) Relative position of the cavity modes for a short optical cavity. Here only one mode lies within the gain bandwidth resulting in a single mode emission. d) Corresponding position of the modes for a long cavity for which several modes lie within the gain bandwidth.

where κ is the cavity decay rate, $\gamma_\perp$ is the decay rate of atomic polarisation, $\gamma_\parallel$ is the decay rate of population inversion, λ is the pumping parameter, E is the field inside the cavity, D is the population inversion, and P is the atomic polarisation. If we consider the trajectory of the Lorenz strange attractor (see Fig. 1(d)) where in the equivalent laser system the dynamic variables x, y and z are the field amplitude (E), polarisation of the medium (P), and the population inversion (D), a point (x,y,z) circles in one region for a while, but then suddenly jumps into another region, where it moves for a while until it jumps, seemingly randomly, back into the first region, and so on; the trajectory never intersects. For the laser, such behaviour not only requires a cavity with high transmission but also a gain of at least nine times that required to produce lasing [8], making the experimental realisation of such operation rather impracticable for most lasers of this simple type. A notable exception are optically-pumped far-infared molecular lasers which are discussed in the next section.

Consequently attention has been given to alternative, though more complex systems, generally with external control (see below) for which such restrictions are in part relaxed. Investigations here have yielded a wealth of identifiable dynamic instability phenomena some in reasonable agreement with theoretical predictions.

Three-Level Laser

Realisation of instability phenomena in the system prescribed by Haken nevertheless remains especially appealing in view of its fundamental simplicity and recently optically pumped far infrared lasers have been identified [9a] as perhaps the most promising candidates in this regard.

Evidence for Lorentz-type chaos in an NH_3 laser emitting at 81 μm optically pumped by an N_2O laser has been recently reported [9b]. Fig. 3, taken from this work, shows motion typical of a Lorentz system, the spiralling around two centres with random jumps from one centre to the next. The abrupt transition from stable to chaotic emission observed with increasing pump strength provides further support for this interpretation.

Nevertheless, optically pumped systems comprise three levels involving pump and laser transitions with a common level (Fig. 4, inset). The consequent coherent interactions between these fields can considerably modify the shape of the gain distribution (see below) from the Lorenzian profile of a two-level system. Indeed the equivalence of 3- to 2-level schemes is only possible for weak pumping and in molecular systems where the polarisation (de-phasing) decay rate for the pump transition is considerably greater than that for the lasing transition; effectively resulting in incoherent pumping [10] although the latter condition is probably difficult to realise in practice.

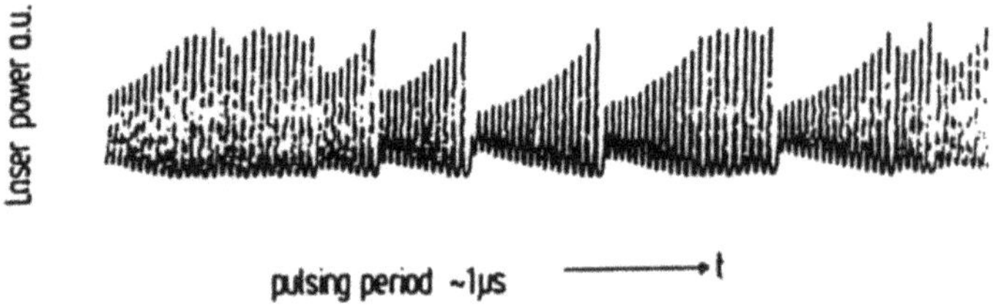

Fig. 3. Lorentz-type chaos in NH_3 laser emitting at 81 μm optically pumped by a N_2O laser. Ratio of laser field decay rate to polarisation decay rate: estimated to be 2. Ratio of pump power to that at first laser threshold 14 [Ref. 9b].

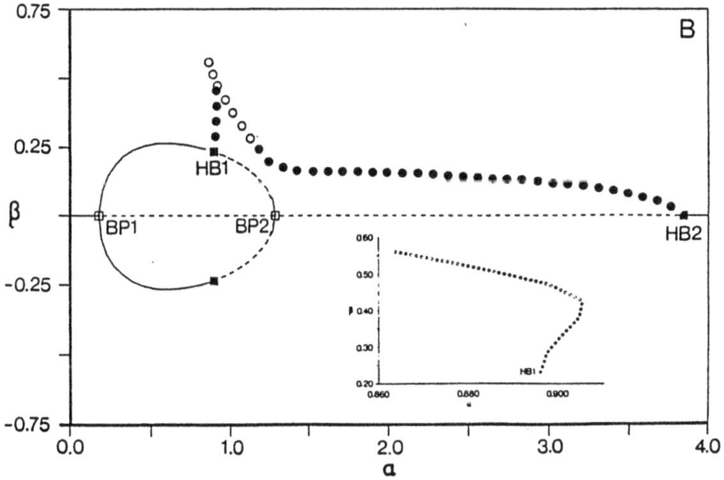

Fig. 4. Bifurcation diagram for laser emission amplitude β at line centre versus pump amplitude α for $\gamma_{\|}/\gamma_{\perp}$ = 0.2. Solid lines denote stable cw operation, dashed lines unstable cw operation, solid dots are branches of stable periodic solutions and open dots branches of unstable periodic solutions. The insets show a schematic of the three-level system under study and a blow-up of the region near HB1. Pitchfork bifurcations BP1 and BP2 in (B) occur at α = .17 and α = 1.29 respectively [Ref. 11].

Recent, more general analysis of line centre, homogeneously broadened three-level systems [11] in which coherent interactions between pump and emission play a central role are discussed below and some results are shown in Figs. 4, 5 and 6. Instabilities and chaos in the emission are found to be sensitive to the relative rates $\gamma_{\|}$ and $\gamma_{\perp}$ and occur close to the normal lasing threshold (compare with nine times above for the 2-level system) and in the bad cavity limit ($\kappa > \gamma_{\|} + \gamma_{\perp}$).

The bifurcation plot of laser emission field amplitude β versus pump field amplitude α (both normalised to $\gamma_{\perp}$) for $\gamma_{\|}/\gamma_{\perp}$ = 0.2 shows a pitchfork bifurcation from a non-lasing to a cw (steady) state (BP1) at α_p = 0.2. With stronger pumping, the lasing emission amplitude initially grows to the point HB1 (Hopf bifurcation point). The periodic solutions emanating from HP1, which are found to be close to the natural relaxation oscillation frequency of the laser are both stable (solid dots) and unstable (open circles); the region $0.905 < \alpha < 1.145$ corresponding to chaotic emission. The blow up of the periodic branch shows, in contrast to the Lorentz model, the bifurcation here to be supercritical with the stable limit cycle losing stability at the limit point (change to open circles).

The narrow region betwen HB1 and this limit point is a preturbulent region where initial random spikings are eventually trapped on the stable limit cycle. Below HB1, a preturbulent regime exists where eventual collapse occurs onto the stable lasing state. The point (α = 1.145) where the chaotic window ends corresponds to the change from unstable to stable limit cycle behaviour (induced by coherent pump-induced Rabi-splitting - see below) on the periodic branch which eventually terminates on the non-lasing branch at α = 3.84 (HB2).

The physical picture of the periodic and chaotic motion can be understood from the plots of the laser output amplitude and the corresponding cavity-dispersion curves versus normalised detuning δ_s of the laser cavity mode from gain line

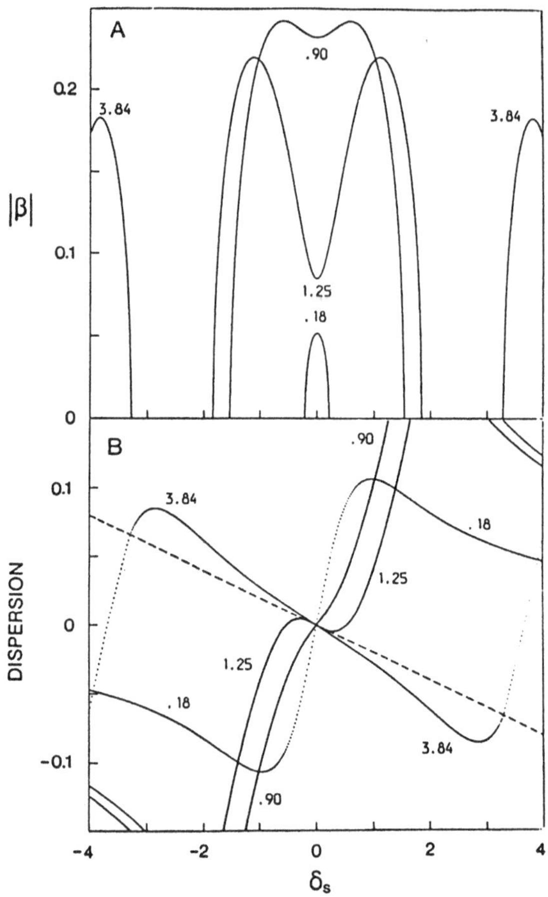

Fig. 5. Plot of the modulus of the laser amplitude β in (A) and cavity-dispersion relation in (B) against laser signal detuning δ_s for b = .2. The laser amplitude β at δ_s = 0 coincides with the branch of steady state solutions in Fig. 4 [Ref. 11].

centre, as shown in Fig. 5. (For single-mode operation the actual intensity at which the laser achieved steady state operation is determined from the intersections of the straight cavity line and the nonlinear dispersion curve). Immediately above threshold at α = .18, the laser emission profile is single-humped and the corresponding dispersion is normal giving line centre emission. The dip and strong anomalous dispersion near line centre at α = 0.9 (HB1 in Fig. 5) result from the coherent effect of pump-induced Rabi splitting of states 1 and 2. At α = 1.25, two new intersections appear with the straight cavity line suggesting the possibility of sideband (Rabi) oscillation [12] consistent with the bifurcation analysis above at α = 1.145. The line centre laser emission drops below threshold at α = 1.29 as shown in Fig. 4. The transition from low amplitude oscillation to non-lasing state at α = 3.84 (HB2 in Fig.4) is accurately captured in Fig. 5 where the double-humped gain region has just shifted outside the cavity-dispersion curve intersection line. The crucial ingredient for chaotic dynamics near threshold in the present model around α ∿ 1.0, is the near coincidence of the induced Rabi sideband frequency arising from the inverted gain distribution and the natural relaxation oscillation frequency of the system. The topology of the attracting set on which the chaotic trajectory lives, involves unstable outward spiralling and inward spiralling associated with motion in the vicinity of the unstable limit cycle (see Fig. 6) manifesting itself in the laser intensity output as a ringing type of oscillation. The unstable outward spiral eventually brings the trajectory close to the unstable saddle point (non-lasing branch) where random delays and switching to the other

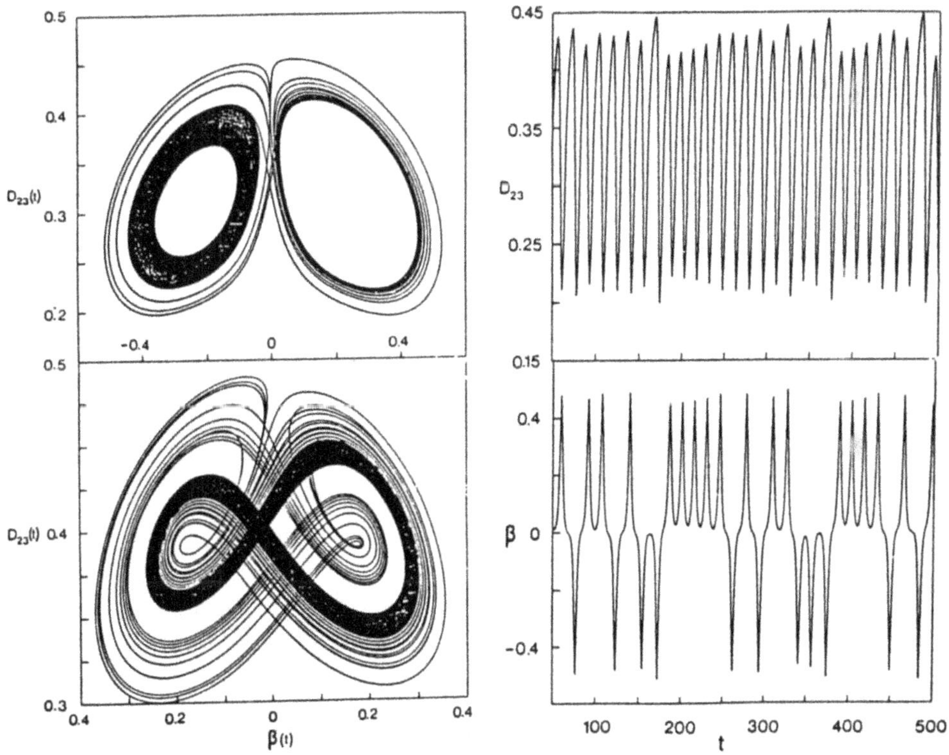

Fig. 6. Trajectories (and corresponding time series) showing graphs of the laser inversion $D_{23}(t)$ versus laser emission amplitude $\beta(t)$ in the preturbulent (top figures and chaotic region (bottom)). The chaotic transient in the preturbulent regime ($\alpha = .903$) locks onto the stable limit cycle in the left lobe ($\beta < 0$ solution) [Ref. 11].

side of the saddle point interrupt the growing ringing-type oscillations associated with unstable outward spirals. The phase portraits and time series shown in Fig. 6 illustrate this type of motion both in the preturbulent ($\alpha = .903$) and chaotic regimes ($\alpha = 1.0$).

Lasers with external control

If one of the variables relaxes faster than the others the stationary solution for that of the laser variable may be taken, so resulting in a reduced number of coupled differential equations; commonly termed adiabatic elimination of the fast variables [3,4]. In many systems, polarisation and population inversion have relaxation times much shorter than the cavity lifetime and both variables can be adiabatically eliminated. With just one variable describing the dynamics, the laser must show a stable behaviour (fixed point in phase space; see Fig. 1(a)). This group of lasers comprises many common systems such as He-Ne, Ar^+, and dye lasers. In some cases, only polarisation is fast and hence two variables describe the dynamics. In this class, we find ruby, Nd and CO_2 lasers which exhibit oscillating behaviour in some conditions, although ringing is always damped.

Since many lasers are not described by the full set of Maxwell-Bloch equations normally chaotic behaviour from these systems cannot be obtained. For these systems with less than three variables the addition of independent external control parameters to the system have been extensively considered [13] as a means to provide the extra degrees of freedom. Active modulation of a parameter such as population inversion, field or cavity length as well as injection of a constant field detuned from the cavity resonance and also the use of intracavity saturable absorbers have all been considered [13-16]. For multimode rather than single-mode lasers intrinsic modulation of inversion (or photon flux) by multimode parametric interaction ensures additional degrees of freedom [13]. When the field is detuned from gain centre the field amplitude, polarisation and population inversion are complex, providing (in the absence of adiabatic elimination) five rather than three nonlinear equations for single-mode systems which is more than sufficient to yield deterministic chaos for suitable parameter values [17]. Also of significance is the remarkably low threshold found for the generation of instabilities and chaos in single-mode inhomogeneously broadened laser systems [18-20]. Compared with homogeneously-broadened systems this is attributed to the increased number of independent gain packets available in inhomogeneous systems. Pulsating instabilities and routes to chaos have also been reported for Raman lasers [21] where the instability threshold is again found to be reduced. Significantly, it is also found that instabilities are greatest in conditions for which the laser produces maximum output.

4. Conclusions

Nonlinear optics,in particular lasers,is proving valuable to the field of nonlinear dynamics and deterministic chaos providing simple optical systems which exhibit the most interesting classes of chaotic behaviour. On the other hand, lasers and related nonlinear optical devices have a large and growing technical application, and the understanding, control and possible exploitation of sources of instability in these systems has considerable practical importance.

Experimental findings are, in general, not yet sufficiently comprehensive to permit the quantitative analysis necessary to fully test the theoretical models. More carefully controlled experiments are forthcoming from which, along with time series, power spectra and identification of routes to chaos, embedding procedures may be implemented to the attractors describing the dynamical behaviour of these systems.

More detailed discussions of points in this review can be found elsewhere. For laser instabilities see refs. [22], for both laser and passive systems (see ref. [23]. Comprehensive treatments on the more general principles of deterministic chaos can be found in refs. [3,4,24].

References

1. E.N. Lorenz, J. Atoms. Sci., 20, 130 (1963)
2. H. Whitney, Annals. Math., 37, 645 (1936).
3. H. Haken, Synergetics - An Introduction (Springer, Berlin, 1983).
4. H. Haken, Advanced Synergetics (Springer, Berlin, 1983).
5. F.J. Feigenbaum, J. Stat. Phys., 19, 25 (1978).
6. P. Manneville and Y. Pomeau, Phys. Lett., 75A, 1 (1979).
7. D. Ruelle and F. Takens, Commun. Math. Phys., 20, 167 (1971).
8. H. Haken, Phys. Lett., 53A, 77 (1975).

9a. C.O. Weiss, W. Klische, Opt. Commun., 5, 47 (1984).

9b. C.O. Weiss and J. Brock, Phys. Rev. Lett., 57, 2804 (12986).

10. M.A. Dupertuis, R.R.E. Salomaa and M.R. Siegrist, Opt. Commun., 57, 410 (1986).
11. J.V. Moloney, J.S. Uppal and R.G. Harrison, Phys. Rev. Lett. (submitted).
12. S.C. Mehendale and R.G. Harrison, Phys. Rev., 34, 1613 (1986).
13. H.J. Scholz, T. Yamada, H. Brant and R. Graham, Phys. Lett., 82A, 321 (1981).

14. F.T. Arecchi, R. Meucci, G.P. Puccioni and J.R. Tredicce, J.R. Phys. Rev. Lett., 49, 1217 (1982).
15. T. Midavaine, D. Dangisse and P. Glorieux, Phys. Rev. Lett., 55, 1989 (1985).
16. L.A. Lugiato, L. Narducci, D.K. Bandy and C.A. Pennise, Opt. Commun., 46, 64 (1983).
17. P. Mandel and H. Zeglache, Opt. Commun., 47, 146 (1983).
18. L.W. Casperson, Phys. Rev. A, 21, 911 (1980).
19. N.B. Abraham et al., Lecture Notes in Physics, Vol. 182, 107 (Springer, Berlin, 1983).
20. R.S. Gioggia and N.B. Abraham, Phys. Rev. Let., 51, 650 (1983).
21. R.G. Harrison and D.J. Biswas, Phys. Rev. Lett., 55, 63 (1985).
22. R.G. Harrison and D.J. Biswas, Progress in Quantum Electronics, 10, 147 (Pergamon, Oxford, 1985).
23. F.T. Arecchi and R.G. Harrison (eds. Instabilities and Chaos in Quantum Optics (Springer, Berlin, in press).
24. H.G. Schuster, Deterministic Chaos (Physik, Berlin, 1984).

Part VI

General Computational Processes

Systems with Statistically Coupled Processors

V. Černý

Institute of Physics and Biophysics, Comenius University,
Mlynská dolina, 842 15 Bratislava, Czechoslovakia

1. Introduction

Many practical problems can be represented by schemes like that in Fig.1. The nodes represent the fact that some local problems have to be solved consistently with the constraints which are represented by the links. The scheme as a whole represents one global problem. The scheme just corresponds to the fact that this global problem has to do with data having rich structure.

A standard "algorithmical" way of solving the global problem is to perform an intelligent analysis and to find an algorithm which solves the problem essentially by performing a linear sequence of algorithmic steps. This one-dimensional logical approach is rather unnatural with respect to the complicated and rich structure of the data. An alternative approach is the distributive computing. The aim is to spread the computing power into the data structure and to solve the global problem by a cooperative effort of many local processes. One can imagine a large transport problem being solved by local dispatchers making their choice of where to send their trucks just on the basis of local information.

Such an approach is not trivial. One can imagine a dispatcher A who would like to know the decision of his neighbour B before he makes his own decision, but the same might be true for the dispatcher B. The problems A and B are interrelated and cannot be trivially solved in parallel.

In the paper /1/ we have proposed a fairly general scheme for controlling multiprocessor networks enabling one to perform interrelated operations in parallel. The basic idea is to substitute strict deterministic constraints between the processors by constraints which are statistical in character. In this way we arrive at a multiprocessor system with statistically coupled processors.

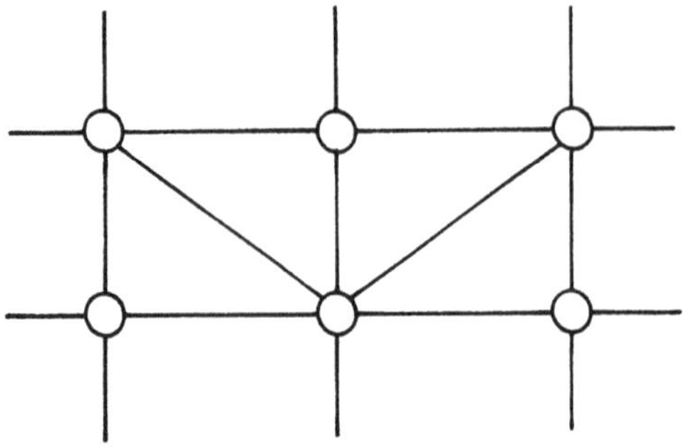

Fig. 1

We intuitively feel that the parallel processing of interrelated operations is widely used in nature. An athlete performing the high jump makes simultaneously complicated movements with different parts of his body. These movements are certainly not independent but they are performed and controlled in parallel. We think that it is possible because the body is elastic enough to allow for certain incompatibilities of the different movements and for quick iterative corrections which help to maintain the incompatibilities at an acceptable level.

Man's ability of having global insight into the nature of complicated problems is - in our view - also based on something like the parallel processing of interrelated operations. For example, the search for the directions of mountain ranges on a map is based on matching results of parallel local observations. Those local observations are not independent. The local situation is perceived and evaluated with respect to what has been found in the neighbouring regions. Here also a certain level of "elasticity" in matching various pieces of information is important. One first gets a rough unfocused picture which is then gradually improved.

In this contribution we first present a simple pedagogical example in which we demonstrate the basic ideas of the method of statistically coupled processors. Then we shall illustrate how the idea of the preceding paragraph - matching interrelated local observations - can be practically realized in an example from picture processing.

2. Distributive Computing as an Optimization Problem

Qualitatively, the method of transforming a global problem of the type described by Fig.1 into a working distributive computing scheme via the method of statistically coupled processors can be summarized in the following three steps.

Step 1. Transform your global problem into an optimization problem by allowing inconsistencies between the local processes: you simply allow local solutions which do not satisfy all the constraints. You have to introduce a measure to quantify the level of inconsistencies. Then what you have to do is to search for such local solutions which minimize the level of inconsistencies.

Step 2. Construct - as a Gedankenexperiment - a would-be analog computer which can solve your optimization problem. By this we mean a physical system the energy of which is equal to the cost function of your optimization problem and the ground state of which is the solution you are looking for.

Step 3. Find the ground state of your analog computer by heating it first to a reasonably high temperature and cooling it slowly to $T = 0$ to avoid local minimum traps which you have probably formed in the Step 1. This is so-called simulated annealing optimization technique /2/, /3/. Of course, you are not expected to perform the analog computing: instead of that you have to simulate the behaviour of the analog physical system on a digital computer. The use of the simulated annealing method allows to treat different degrees of freedom - that means different local decision problems - in parallel. Therefore you can use the naive multiprocessor network you have started with; however, now your "local dispatchers" will have a well-defined strategy of how to proceed to arrive at a globally acceptable solution.

3. A Simple Example: Two Coupled Equations

In this section we shall illustrate the general rules for the construction of the system with statistically coupled processors on a trivial example: two processors operating in parallel solving the system of two coupled linear equations

$$\begin{aligned} x + y &= a \\ x - y &= b . \end{aligned} \tag{1}$$

This is a problem the global algorithmic solution of which has been known for ages. Deterministic iterative schemes for solving systems of linear equations are also well known and many of them can easily be parallelized. The method of statistically coupled processors will certainly be worse than those standard methods for this case, therefore this example has only a pedagogical value.

Let us imagine a processor A which can solve the first equation for y at a given fixed x and a second processor B which can solve the second equation for x at a given fixed y. The constraints between these two processes are shown in Fig.2: the value x at the input of A should be the same as the value at the output of B and the value at the input of B should be the same as the value at the output of A. Let us stress that the naive network based on a simple feed-back as shown in Fig.2 would not work: it will oscillate.

We shall now follow the rules described in the Section 2.

Step 1. We first decouple the two processes by violating the constraints allowing different values at the inputs and outputs and we arrive at a situation shown in Fig.3. The state of the two-processor system is defined by two numbers x' and y' and we easily introduce a measure for the level of inconsistency

$$E(x',y') = (x - x')^2 + (y - y')^2 . \tag{2}$$

Now we have to look for the state (x',y') for which $E(x',y')$ is minimal. The minimal value of $E(x',y')$ is, of course, equal to zero and the corresponding state is the solution of the equations(1) .

Step 2. Now we shall construct an analog device which can solve our optimization problem. We take two scales, two springs, some

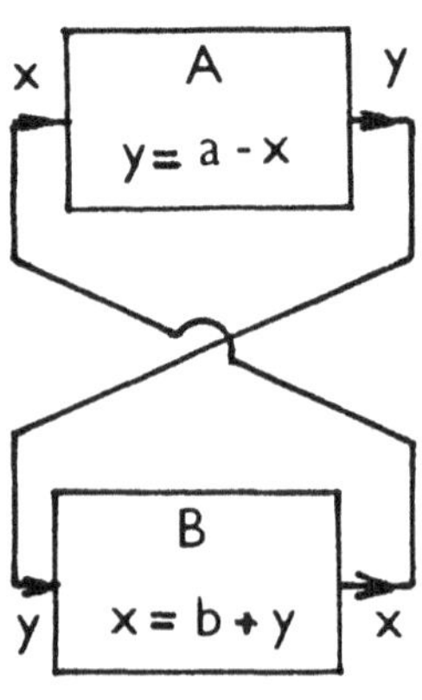

Fig. 2

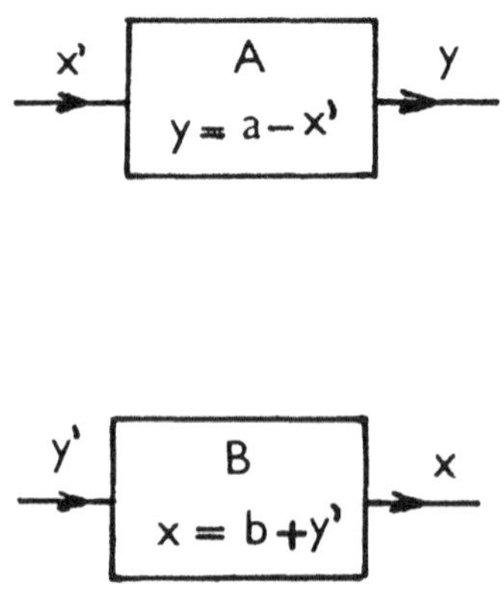

Fig. 3

Fig. 4

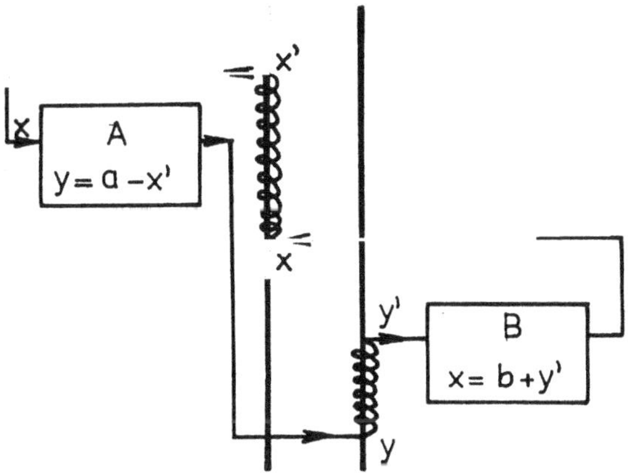

read-out devices and some servomechanism devices and we connect all this to the processors A and B as it is shown in Fig.4. The device works as follows. The position x' of the top end of the first spring is read and the value x' is put into the input of the processor A. This processor immediately calculates the value $y = a - x'$ and by means of a servomechanism sets the bottom end of the second spring into the position y. The position y' of the top end of the second spring is read and fed into the input of the processor B. This processor calculates the value $x = b + y'$ and sets the bottom end of the first spring into the position x. Since the processors operate much faster then the response time of the springs, we have got an analog device with only two degrees of freedom x' and y' and the energy of the device in the state x',y' is given by the expression (2).

Step 3. Now we have to heat the system and cool it down to zero temperature and the device will - by its spontaneous dynamics - arrive at its ground state, which is given by the solution of the equations (1). The springs will simply shrink to zero length, but this is possible only in special positions x' and y' given by the solution of the problem. Since the energy is given by a quadratic expression, there are no local minima and the annealing procedure is not necessary. In more complicated problems we have to perform the annealing in order to arrive at the global minimum.

We shall proceed by describing the method of simulation of our fictitious analog device by means of a simple two-processor network. The method is a parallelized version of the simulated annealing method of optimization /2/,/3/. Since there are review papers available which discuss the simulated annealing method in detail /4/,/5/, we skip a general discussion of the method here. We present only the resulting algorithm according to which the processors in our two-processor network should operate.

The processor A will operate according to the following algorithm.

Step 3.1. Ask the processor B what is its current meaning about its input and output values y' and x respectively.

Step 3.2. Generate randomly a trial value $\tilde{x}'$ in the neighbourhood of what is your current meaning about the input value x'. Calculate the corresponding output value $\tilde{y} = a - \tilde{x}'$. The output value corresponding to the current value x' is denoted as y.

Step 3.3. Evaluate the current energy E and the trial energy $\tilde{E}$ according to the expressions

$$E = (y - y')^2 + (x' - x)^2$$

$$\tilde{E} = (\tilde{y} - y')^2 + (\tilde{x}' - x)^2$$

where y' and x are the values provided by the processor B.

Step 3.4. Now decide whether to accept the trial state $(\tilde{x}', \tilde{y})$ as a new current state(x',y) or to forget about this trial and to make a new trial in the neighbourhood of the unchanged current state (x',y). The decision is to be made according to the following rule:

If $\tilde{E} \leqslant E$ then accept the trial state $(\tilde{x}', \tilde{y})$ as a new current state (x',y) and go to the Step 3.1 with this new current state.

If $\tilde{E} > E$ then generate a random number $t \in (0, 1)$ and if

$$t < \exp(-(\tilde{E} - E)/T)$$

then accept the trial state$(\tilde{x}', \tilde{y})$ as a new current state (x',y) and go to the Step 3.1 with this new current state, or else go to the Step 3.1 with unchanged current state(x',y)

A few comments are in order.

Comment 1. The core of the algorithm is in the Step 3.4 and it is clear that it is the well-known Metropolis algorithm /6/ simulating a system in contact with a thermostat at a temperature T.

Comment 2. The processor B will operate according to a similar algorithm with the rôle of x and x' and y and y' interchanged. The temperature T should be the same for both the processors.

Comment 3. The algorithm as described by Step 3.1 to Step 3.4 is an infinite loop, it produces an infinite sequence of current states (x',y). It has to be started in a randomly chosen initial state. After the start, the temperature T should be gradually decreased to zero and a suitable stopping procedure should be used to terminate the process when no further improvements in the energy are observed. The specific way of how to decrease the temperature - the so-called cooling schedule - is a delicate question which has been extensively discussed in the literature /4/,/5/ and we shall not elaborate on this point here.

Comment 4. A detailed discussion of what is meant by the neighbourhood of the value x' in the Step 3.2 can be found in /7/.

Based on this specific simple example one can now easily formulate the algorithm for a network of statistically coupled processors in more complicated cases. The main trick is that a processor does not directly use the output values of its neighbours as its input but rather tries to guess what the correct input value should be and evaluates the correctness of its guess according to the Metropolis rule. The scheme is fairly general, and the intuition based on experience with statistical physics problems suggests that the iterative scheme obtained in this way should be convergent.

In general, the convergence of the simulated annealing optimization method can be proved under very weak assumptions about the problem. The question of the convergence **rate**, however, is a delicate one. Although the theory does provide a few answers, a numerical experiment is still necessary to test the applicability of the method for every new class of problems.

In the next section we present a particular example from a large class of pattern recognition problems for which the method of statistically coupled processors seems to be very promising /8/,/9/.

4 Picture processing by statistically coupled processors

Real-time picture processing has been recently receiving much attention with respect to various parallel computational schemes and computer architectures. It is quite natural: picture processing means manipulations with large amounts of data in a few hundredths of a second. However, picture processing is a complex task. Usually it means a hierarchy of processing starting with noise reduction and picture enhancement, through edge detection and segmentation to feature extraction to high level processes like syntactical analysis and interpretation.

The low-level processes can usually be easily parallelized: they are applied locally and the operations in neighbouring regions are not interrelated. However, the high-level operations are more complicated and they work with the picture content from different locations on the picture - essentially they are looking for correlations between various parts of the picture.

In this section we shall demonstrate with a simple but realistic example that the method of statistically coupled processors can be used to parallelize higher level operations as well. Moreover, we shall show that it is possible to combine the low-level and the higher level processes into a single process and therefore a preliminary information from the higher level operations can be used to make the low-level operations more reliable. For example the preliminary information on edge detection and segmentation will be used to guide the process of noise reduction so that the edges and segments will be preserved during the filtration process.

Our example will demonstrate the combined operation of noise reduction, edge detection and segmentation using a massively parallel network of statistically coupled processors. We shall work with binary pictures with pixels having hexagonal geometry: the pixels will be triangles arranged into a hexagonal honeycomb-like structure. The information from six pixels forming an elementary hexagon will be processed by one processor. A possible subpicture seen by one processor is shown in Fig.5. We shall consider 127 processors forming a hexagonal array presented in Fig.6 where an object - a circle - is projected on the screen. In real situations one should expect pictures distorted by some noise signal so that a real picture looks like Fig.7.

Our aim now is to use the processor array to segment the objects in pictures like that in Fig.7. The shape and the positions of the objects are a priori not known.

If the situation was like that in Fig.6 the solution of the problem would be simple. The processors which would see their hexagonal field completely white would claim that they sit outside the object, the processors which would see their field partly black and partly white would claim that they sit on the edge of the object, and the

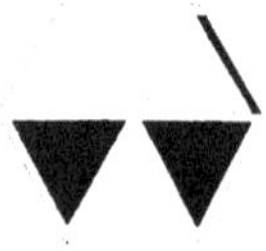

Fig. 5

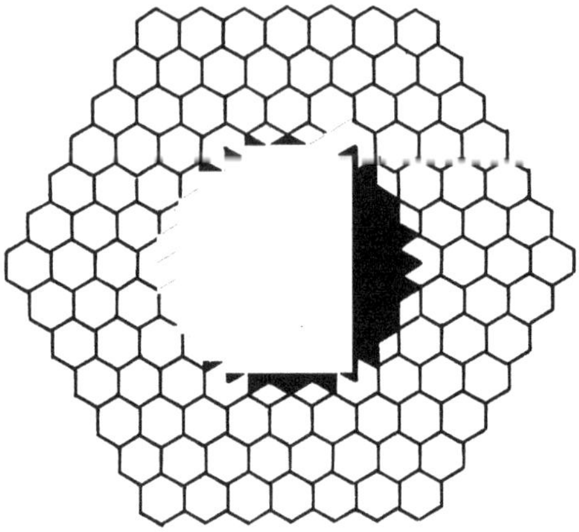

Fig. 6

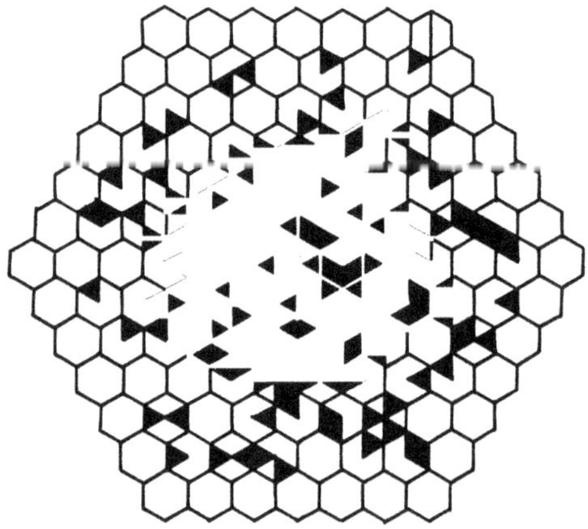

Fig. 7

processors which would see their field completely black would claim that they sit inside the object.

In the presence of noise the algorithm is more complicated. We describe the solution of the problem by constructing a physical system the ground state of which will provide the required result. It means that in the notation of the previous section we skip the Step 1 and we perform directly the Step 2 without trying to define the problem in a more exact mathematical language.

We consider a hexagonal spin lattice with nearest neighbour interaction. We assume that each spin can be in one of the eight possible states. A certain template corresponds to each of these states; the templates are presented in Fig.8. The picture which should be processed will represent an external field for our spin system. The interaction energy of the spins with this external field will be calculated by counting the number of pixels in the templates corresponding to the states considered which differ from the same pixels in the picture.

In the situation without the noise we could use the spin system without interaction between spins to segment the objects in the picture: the ground state of the spin system would correspond to the best possible matching between the templates and the picture.

In the presence of noise the best possible local matching does not necessarily mean the best global solution: fluctuations of the noise signal may distort the picture in such a way that the local interpretation would be wrong. To overcome this difficulty one has to consider the correlations between local interpretations. Here the idea of statistical coupling between the processors is essential.

One can easily define consistency requirements between the local decisions as for example that a processor cannot claim that it is

State	Template	Meaning
0		Outside the object.
1		Edge. Object expected to be in the direction 1.
2		Edge. Object expected to be in the direction 2.
3		Edge. Object expected to be in the direction 3.
4		Edge. Object expected to be in the direction 4.
5		Edge. Object expected to be in the direction 5.
6		Edge. Object expected to be in the direction 6.
7		Inside the object.

Fig. 8

outside the object if all his neighbours say they are inside the object. The consistency conditions are schematically presented in Fig.9.

We can now introduce the nearest neighbour interaction into our spin model which takes into account the consistency requirements. The complete expression for the energy of a particular spin in the state h will be

$$E(h) = n(h) - \ln(T/b)\, m(h) \qquad (3)$$

where n(h) is the number of pixels in the template h different from the corresponding pixels in the picture and m(h) is the number of neighbouring spins the state of which is inconsistent with the state h of the spin considered. Thus the first term measures the energy of interaction with the external field, the second term is the interaction energy between the neighbouring spins.

The coefficient between the second term in (3) depends on the temperature T; b is a constant of the order of the initial annealing temperature. Before discussing the rationale for this choice let us investigate how the spin system works.

At high temperatures the second term in (3) is not important and the energy of the states is given by the degree of the local matching of the templates to the picture. When we start to cool the system, the interaction between the spins becomes important and in the limit of zero temperatures only consistent states are effectively permitted by the energy.

The reason for having the energy dependent on the temperature lies in the fact that we have to do with two different optimization problems. The first one is the original problem of the best possible

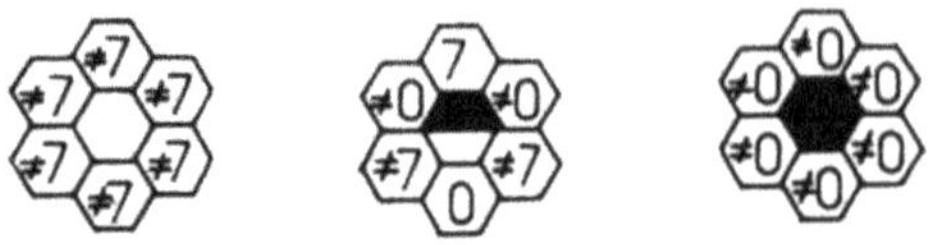

Fig. 9

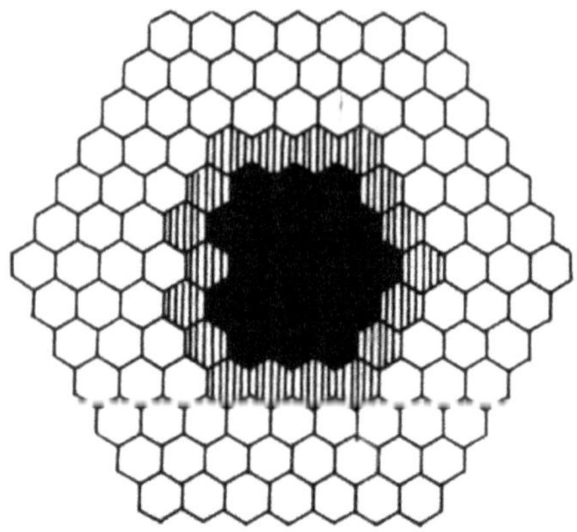

Fig. 10

matching of the templates, the second one has arisen through the method of statistically coupled processors as a substitution for the rigid consistency requirements. The constrained optimization is changed into the unconstrained optimizations within a large superimposed optimization problem which takes into account the constraints. In order not to have optimization in optimization but rather a single large optimization problem, we have developed the method of the energy dependent on the temperature.

It is clear how the behaviour of our fictitious spin system can be simulated by a multiprocessor network using the Metropolis algorithm. The result of a particular computer simulation is presented in Fig.10. The processors which ended in the state 0 - "outside"- are white, the processors which ended in states 1 to 6 - "edge" are cross-hatched, and the processors which ended in the state 7 "inside" are black. The result corresponds to the picture presented in Fig.7, and was obtained after about 100 iterations per processor. One hundred Monte Carlo trials per processor is a reasonably small number, and since this relaxation time needed to segment one object on a small screen is the same as the time needed to segment many objects on a large screen, the method might have practical significance when massively parallel multiprocessor networks become cost-effective.

5. Conclusions

We have shown in two simple examples how the idea of statistically coupled processors works. We are now working on more complicated but practically oriented problems from picture processing and the first results seem to be encouraging. Other people have used a very similar idea to investigate a scheme for an associative memory /10/.

We believe there are many areas where our scheme might be useful. As already said in the introduction there is an obvious difference between a rigid industrial robot and a natural system like Man. A robot of today playing table tennis would try to solve perhaps some complicated systems of equations while a man does a great part of the job by reflexes controlled from many local processing centres. We believe the model of statistically coupled processors may be useful to study questions of how a coordinated behaviour arises from parallel quasiindependent activities. Some authors even argue /10/

that the parallel with the natural systems might be closer than one would perhaps think. It is well known that there are noise signals in the neural networks. It may well be that this noise is not just an imperfection of Nature. The noise might be functional: a probabilistic logic of the "local decisions" as required by the scheme of statistically coupled processors can easily be implemented using such a noise.

References

1. V.Černý: Multiprocessor System as a Statistical Ensemble, a Way Towards General Purpose Parallel Processing and MIMD Computers, preprint, Comenius Univ. Bratislava, 1983
2. S.Kirkpatrick,C.D.Gelat,M.P.Vecchi, Science 220, 671 1983
3. V.Černý, A Thermodynamical Approach to the Travelling Salesman Problem, preprint, Comenius Univ. Bratislava, 1982 and J.of Optim.Theory and Appl. 45, 41 1985
4. L.T.Wille, Optimization by Simulated Annealing, an Overview and Some Case Studies, In Inf. Quarterly for Comp. Simul. of Condensed Physes, Daresbury Lab., No. 20, 1986
5. E.H.L.Aarts, P.J.M. van Laarhoven: Simulated Annealing, Theory and Applications, to be published
6. N.Metropolis et al.: J.Chem.Phys. 21, 1067 1953
7. V.Černý: preprint Helsinki Univ. HU-TFT-84-51, 1984
8. V.Černý: Annealing Algorithm and Parallel Processing, an Option to Approach Pattern Recognition Problems, preprint, Comenius Univ. Bratislava, 1984
9. V.Černý: Picture Processing by Statistically Coupled Processors, Relaxation Syntactical Analysis, IBM Europe Summer School, Oberlech July 1986 and preprint, Comenius Univ. Bratislava, 1986
10. G.F.Hinton et al.: preprint Carnegie-Mellon Univ. CMU-CS-84-119 1984

Chaotic Dynamical Systems as Machines

J.L. McCauley

Institute for Energy Technology, P.O. Box 40,
N-2007 Kjeller, Norway*

1. Preliminary Discussion

The subject of this discussion belongs to no single well-established field but overlaps physics, mathematics and computer science. It can begin with a rather simple question, "when can a chaotic dynamical system be regarded as a machine, or as a model of a machine?" Also, "when is the output (behaviour) of a machine chaotic?" The two questions are related and one is led to them by asking, "to what extent is the orbit of a deterministic but chaotic dynamical system computable?" and, "in what sense is such a computed orbit chaotic?" In the past, appeals have been made to symbolic dynamics and to algorithmic complexity, but those efforts have not resolved these questions. The systems that we shall concentrate upon are purely deterministic, so that the chaos must be generated entirely by the dynamical system during the computation, and not by the effect of any external noise.

We shall also address a related question, namely, does the β-shadowing lemma have any implication for computation? In particular, does it justify the use of floating-point or other equally bad arithmetic in the study of chaotic attractors or mixing systems? Here, we have in mind any computation where the precision and truncation/roundoff decisions are fixed arbitrarily in advance, with no regard for the demands made by the effect of a positive Liapunov exponent. One purpose of this lecture is to emphasize and make clear the nature of those demands and to indicate how they can be realized by the use of algorithms for computable irrational numbers. In this way are we led to a clearer and more basic view of the limitations on predictability in deterministic chaotic dynamics, devoid of the misleading mythology that ordinary truncation/roundoff decisions simulate either randomness or real physical perturbations during a machine computation. The main effect of ordinary truncation and roundoff is in fact to impose periodicity, even where none exists for the dynamical system under consideration [1]. In fact, it is necessary to view a floating-point computation as a pure numerical experiment, unjustified by any current-day mathematics, and distinct from "real" experiments on natural chaotic dynamical systems. Toward these ends, let us begin with the computer scientists' definition of a computation.

2. The Mechanical Nature of Computation

It is useful to define an automaton (processor) as a machine or a mathematical model of a machine, and to realize from the start that the discretization of any dynamical system for the purpose of computation amounts to the replacement of some natural system by an "artificial automaton" [2]. This is a necessary replacement in the study of complicated systems, and the goal is to make an accurate enough replacement so that both systems generate essentially the same statistics from their "orbits". We concentrate upon single orbits from well-defined initial conditions, for it is a single typical or "generic" orbit whose closure defines the attractor or invariant density in a chaotic dynamical system. So, for computation, the natural system or continuum set of equations is replaced by the behaviour of machine, something purely mechanical and predictable. One cannot help but be reminded of Boltzmann's early efforts to understand entropy and the Second Law via purely mechanical models, involving gears and so on.

The machine or its model has a finite number of internal states and the computation is carried out by transitions among these states. Given a definite input x,a unique output x' follows. For example, with $x' = Dx(1-x)$, the internal state transitions amount to constructing $1 - x$ from x, multiplication by x and then by D. However, a severe restriction is absolutely necessary: x and x' must be either symbols or finite-length strings of symbols in some finite alphabet. For example, with base μ arithmetic, we can form $x = 0.\varepsilon_1\varepsilon_2\ldots.\varepsilon_N$ $x' = 0.\varepsilon_1\varepsilon_2\ldots.\varepsilon_{N'}$, where ε_i and ε_j take on integer values 0, 1, 2, 3,..., μ-1. Furthermore, the internal state transitions are finite in number and only finite-length words are permitted therein, as well [3]. So, in computation, we have the picture of a machine receiving a symbol string of length N and producing as a consequence a symbol string of length N'. The procedure is purely mechanical and there is no "randomness" at any stage of the operation. This is the model of a machine that we should bear in mind in all that follows.

For the purpose of pedagogy, let us consider a one-dimensional iterated map of the unit interval

$$x_{n+1} = f(x_n, D) \qquad (1)$$

with control parameter D. An orbit is given by $x \to f(x,D) \to f^{(2)}(x,D)$.. and in order to carry out the replacement of (1) by an automaton systematically, we need only discretize x_n and D in some integral base μ (= 2, 3, 4...) of arithmetic,

$$x_n = \sum_{i=1}^{N(n)} \varepsilon_i(n)/\mu^i, \qquad D = \sum_{j=-M}^{M} \delta_j/\mu^j \qquad (2)$$

where the expansion coefficients $\varepsilon_i(n)$ and δ_j take an on integral values 0,1,2,... μ-1. At stage n, we then have a natural partitioning of the phase space into $\mu^{N(n)}$ bins, each of width $\mu^{-N(n)}$, and the computation of (1) then amounts to dropping balls into these bins with a definite time-sequence, thus building up a histogram that reflects the automaton's statistics for a given initial condition $x_0 = 0.\varepsilon_1(0)\varepsilon_2(0)\ldots\varepsilon_{N(0)}(0)$. <u>If</u> we can choose a "generic" initial condition and carry out the computation correctly, <u>then</u> the large n limit would yield the correct statistics of the chaotic system. However, that is not what is done in a typical computation. One central problem is that, for chaotic maps, the precision N(n) cannot be fixed in advance, but must be permitted to increase with n according to the demand of a positive Liapunov exponent. To explain this, it is useful to restrict ourselves to "real arithmetic", (fixed-point operations [4]), for floating-point arithmetic introduces errors that cannot easily be controlled, especially when the system is chaotic.

In the analytic formulation of chaos theory, a positive Liapunov exponent λ leads to exponentially fast magnification of any error or change in the initial condition x_0 of (1):

$$|\delta x_n| \sim e^{n\lambda}\, |\delta x_0|. \qquad (3)$$

In the language of automata, we should use $|\delta x_0| \sim \mu^{-N}$ and $|\delta x_n| \sim 1$ to obtain

$$n \sim N \ln\mu/\lambda \qquad (4a)$$

as the average number of iterations that can be performed before an error in the least significant digit $\varepsilon_N(0)$ of x_0 is propagated to the left to become an error in $\varepsilon_1(n)$, the leading digit of x_n. That is, $\lambda > 0$ represents an information flow from right to left in these strings as n increases. This point is central for understanding the meaning of β-shadowing in the context of machine computation [6]. Geometrically, this finite average information flow rate is due to "hyperbolic" regions in phase space, namely, regions where $|f'| > 1$, so that f^{-1} contracts lengths. For the logistic map $f(x,D) = Dx(1-x)$, these regions fall near the end points at $[0,(1-1/D)/2)$ and $((1+1/D)/2,1]$. When $D = 4$, this region occupies 7/8 of the phase space and the logistic map is then conjugate to a purely hyperbolic map, the tent map [7]. The logistic map is not hyperbolic, for f^{-1} is not everywhere contracting over the set [0,1].

However, with $D = 4$, almost all orbits of $f(x,D) = 4x(1-x)$ are chaotic and even mixing [8], and since chaos occurs with measure one, there is the tendency to expect to see it in computation. A simple counterexample shows that this argument is too simple: consider the orbit of the $D = 4$ logistic map when $x_0 = 1/8$ and with $\mu = 2$. $x_0 = 1/8 = 0.001$ is the shortest binary string that does not lead directly to an unstable fixed point upon iteration of the map. Consider the following automaton: we use precision $N = 6$, so that all strings x_n are of 6-bit length, and to be realistic we shall perform the multiplication $x_n*(1-x_n)$ on a twelve-bit multiplication register and then truncate to the first six bits to obtain x_n, after shifting the binary point two places to the right. The phase space is here partitioned into $2^6 = 64$ bins, but after exactly 6 iterations of the map we are stuck on the four cycle 0.111111 → 0.000011 → 0.001011 → .0.100100 → 0.111111. If you try different initial conditions, and different precision, you will find that this behaviour is rather typical of the logistic automaton. Related observations were made by using a different model of a computer [9]. The tendency is to get stuck on a short, rather than on a long period pseudo-orbit. These computed orbits are called pseudo-orbits because they are not true orbits of the dynamical system. For $x_0 = 1/8$, the logistic map's orbit is non-periodic and perhaps has a closure that generates the well-known invariant density $p(x) = (x(1-x))^{-1}/\pi$. With the pseudo-orbit, in contrast, we see nothing but "truncation-noise" after $n \backsim N\ln 2/\lambda = N$ iterations when $\mu = 2$. von Neumann observed over 30 years ago that the logistic map would serve as a good pseudo-random number generator were it not for the fact that one sees nothing but roundoff noise after only a few iterations on a machine. By looking at the pseudo-orbit for $x_0 = 1/8$ computed above, and comparing with the histogram predicted by the known invariant density, we see that the pseudo-orbit produces correct bin statistics only for a partitioning of the phase space [0,1] into 2 bins of equal size, but not for a partitioning into 4 or more bins of equal size. In general, we expect to find that

$$2^{N_b} \ll 2^{N_p} \ll 2^N \tag{4b}$$

where N_b is the precision of the partioning for which correct statistics are obtained, $\exp N_p \ln 2$ is the cycle-length of the pseudo-orbit and N is the maximum precision in the computation. We do not expect qualitatively different behaviour if floating-point arithmetic is used instead of fixed-point operations.

3. The β-shadowing Lemma in Computation

Consider an iteration of (1) where a bounded error is made at each stage n of the computation. The inequality

$$|x_{n+1} - f(x_n,D)| < \alpha \tag{5}$$

for some range of n defines an α-pseudo-orbit $\{x_n\}$. This pseudo-orbit of f is in fact an exact orbit of a different function constructed by the machine that performs the computation. Does this orbit generate statistics that at some level of coarse-graining agree with those generated by (1)? The β-shadowing lemma has been advanced as a reason why this should be the case [10,11]. A true orbit $\{y_n\}$ β-shadows the α-pseudo-orbit if $|y_n - x_n| < \beta$ for the same range of n. It has been stated [11] that one can think of a pseudo-orbit as a trajectory in a random perturbation of f which allows one to perturb f independently at each successive iteration by an amount smaller than α, and that such orbits are essentially like those realized in numerical iterations of maps by a computer. The last part of this assertion is not correct.

Before continuing, let us state the shadowing-lemma [6,11,12]: let Λ be a hyperbolic invariant set of f. Then for every $\beta > 0$, there is an $\alpha > 0$ such that every α-pseudo-orbit in Λ is β-shadowed by a true orbit of f in Λ. This suggests that, for hyberbolic systems, the computed orbit is an approximation to <u>some</u> true orbit of the dynamical system. It has been argued [10] that this lemma justifies the use of ordinary truncation/roundoff methods with fixed-precision arithmetic in the study of chaos. Those authors used linear hyperbolic systems to advance their argument, but we shall use the same system to show by counterexample that the conclusion does not follow. Indeed, machine truncation/roundoff errors are neither independent nor random. They are perfectly deterministic and do not even serve as a good pseudo-random number generator (cf. [4], where it is explained how difficult it is to adequately simulate randomness via machines).

To make the point, let $f(x,D) = Dx \bmod 1$ with $D \geq 2$ an integer. Define a pseudo-orbit of f according to a uniform error/iteration,

$$x_{n+1} = Dx_n + \alpha \bmod 1. \qquad (6)$$

Since $\{x_n\}$ is generated on a machine, all the x_n are rational and since $\alpha = -Dx_0 + x_1$, α also must be rational. If we choose $y_0 = x_0 + \alpha/(D-1)$, then the orbit given by $y_{n+1} = Dy_n \bmod 1$ shadows $\{x_n\}$ with $\beta = \alpha/(D-1)$. Since y_0 is rational, the shadowing orbit is unstable periodic rather than chaotic. It is easy to show that $\{x_n\}$ is periodic, and the conclusion is as follows [6]: periodic pseudo-orbits are at best shadowed by unstable periodic orbits, which means that you cannot use the shadowing lemma to argue that your pseudo-orbit gives correct chaotic attractor statistics at any level of coarse-graining. <u>The shadowing lemma, in the absence of further assumptions, has no implication for computation.</u> Pseudo-orbits may indeed yield correct statistics in some cases at some coarse-grained level, but if so, then shadowing is not the reason for the success. Instead, we must ask, when do long-period (unstable) periodic orbits reflect the statistics of non-periodic orbits, to within some degree of coarse-graining [13,14].

4. Turing, Machines and Chaos

We need not rely upon "bad arithmetic" or any external noise in order to explain deterministic chaos in computation. If we pay attention to Turing's ideas, then we can see how to correctly formulate chaos theory in the context of computation and even arrive at some basic limitations on long-time predictability. Turing observed that (correct) computation is completely mechanical and can just as well be done by a suitable machine, since there is no room for human invention or intervention in the process. Underlying this notion is the fact that digit strings of numbers are fixed once and for all, beyond the influence of the computer. A computable number [15,16] is a number whose "decimal" expansion $\varepsilon_1\varepsilon_2\ldots\varepsilon_N\ldots$ can be generated by an algorithm, for as many digits ε_N as you like. Turing proved that these numbers form only a <u>countable</u> subset of the continuum, so that almost all numbers that can be <u>defined</u> to exist can never be computed by any possible algorithm. All irrationals can be defined as limits of infinite sequences of computable numbers (rationals), but almost all of these limits are non-computable. That, in a nutshell, is why mathematical properties that are true "with measure-one" are not necessarily easily (or ever) reflected in computation; computation takes place on a measure-zero subset. There is no essential difference between functions and numbers in computation [16,17], so we can assert with some degree of confidence that anything that can be computed in chaos theory cannot be more chaotic than sequences constructed from decimal expansions of the computable irrational numbers. $\sqrt{2}$, e, π, and all numbers whose decimals can be computed, e.g. by continued fractions, are computable. In other words, these are the numbers that can be computed by an appropriate automaton, or machine. The fact that all numbers can be <u>defined</u> by continued fractions does not mean that all numbers can be generated via continued fractions; the "seed" necessary to construct the first step in the sequence can be found in <u>at most</u> countably many cases.

These ideas can be used to clarify our notion of deterministic chaos, and to perform better computations. To illustrate this, let $f(x,D) = Dx \bmod 1$ with D an integer ≥ 2. If x_0 is rational, its decimal expansion is periodic in every base $\mu = 2, 3, \ldots,$ and the orbit is periodic. If x_0 is irrational, its decimal expansion is non-periodic in every base μ and the orbit is nonperiodic, hence chaotic, since $\lambda = \ln D > 0$. If x_0 is normal to base D, then the closure of the orbit yields even bin statistics (in the long run) for every degree of coarse-graining, hence the closure of the orbit generates a uniform invariant density $p(x) = 1$ [18]. Almost all numbers are normal to every integral base of arithmetic. The study of normal numbers goes back to Borel [18,19].

Using Turing's ideas, how can we generate true chaos for large but finite n, with N finite? Let D = 3 so that $x_n = 3^n x_0 \bmod 1$, and let $\mu = 2$. Choose, for example, $x_0 = \sqrt{2}-1$, whose binary expansion is

$$\sqrt{2} - 1 = 0.0110101000001001\ldots \qquad (7)$$

to 16 bit accuracy. Let us first evaluate all the x_n to 3-bit accuracy. If we set $x_0 = 0.011$, we obtain the 2-cycle $0.011 \rightarrow 0.001 \rightarrow 0.011 \rightarrow$ .. If, on the other hand, we write down exactly the first three bits of every iterate

$$x_n = 3^n(\sqrt{2}-1) \bmod 1, \qquad (8)$$

then the result is quite different. Since $3^n = (2+1)^n = 2^n + \ldots$, each factor of 2 gives a bit-shift to the right in $\sqrt{2}-1$, so there is a tendency to lose information. However, it is not <u>necessary</u> to lose that information: to compute the first 3 bits in x_2, you need 6-bits in $\sqrt{2}-1$, and so on. The Liapunov exponent $\lambda = \ln 3$ tells you, on the average, how many bits you must keep in x_0 in order to compute every iterate to within at least 3-bits (remember that $n \sim N\ln\mu/\lambda$). The result is the non-periodic pseudo-orbit $.011 \rightarrow .001 \rightarrow .101 \rightarrow \ldots$ which is precisely the 3-bit truncation of the exact chaotic orbit of $\sqrt{2}-1$ [6]! The computed orbit is non-periodic, hence chaotic, because the binary expansion of $\sqrt{2}-1$ is non-periodic. A similar notion can be applied to any computable map f of the unit interval [1] and to flows in phase space defined by computable differential equations [20]. We have also used computable irrational numbers to generate non-periodic orbits of maps via non-standard truncation/roundoff decisions [6].

When maps or differential equations are digitalized for computation, sensitivity to initial data translates into a finite average information flow rate, right to left in digit strings, and this means that <u>computation time</u> becomes the fundamental limitation on long-time prediction in chaotic systems. With the Bernoulli shift $f(x,D) = 3x \bmod 1$ above, you must compute $N \sim n\ln 3/\ln 2$ bits in x_0 in order to know even the first bit of x_n correctly. In the logistic map $f(x,D) = 4x(1-x)$ and with x_0 a finite length binary string, you must know $\sim 2^n = e^{n\ln 2} = e^{n\lambda}$ bits in x_0 in order to correctly predict even the first bit in x_{2n}. In general, to generate symbolic dynamics for large n, you must know many iterates to within a high degree of accuracy. This is the essence of deterministic chaos within the setting of machine computation.

<u>5. Questions of Randomness, Complexity and Cellular Automata</u>

Clearly, there is no randomness after a finite number n of iterations of a map. Randomness can only arise in the strict sense in the limit $n \rightarrow \infty$, which is in every case non-computable. Or, as von Neumann put it, anyone who thinks that he can get randomness from arithmetic is living in a "state of sin" [4].

With $f(x,D) = Dx \bmod 1$, $D \geq 2$ integral, if we choose for x_0 a number that is normal to base D, then we should expect more or less even bin statistics for large but finite n [18]. For $n \to \infty$ this yields "weak randomness". For a definition of "strong randomness", see Knuth [4].

To make use of normality in computation, we must be able to find algorithms that generate normal numbers, and it is well known that these algorithms are hard to find [4,18,19]. There is at least one that is known, namely, write down the natural numbers in any integral base. For $\mu = 3$, the first few digits are

$$0.1210111220212210\ldots \quad (9)$$

It is not known (to the best of the author's knowledge) whether this number is normal to any other base, or even whether $\sqrt{2}$, e or π are normal to any base μ [18]. For a large number of digits, π has been shown to be "effectively normal" to base 10 for the purpose of playing poker [21].

The binary expansions of numbers in the unit interval are formally in one-to-one correspondence with all the fair-coin tossing sequences, as was noted by Borel [19], and almost all of these binary sequences have been proven to be "algorithmically complex" [22]. A binary sequence is called algorithmically complex if the number of bits K_N in the shortest computer program that generates N-bits of the sequence obeys $K_N/N > 0$ as $N \to \infty$. Ford [23] has argued that algorithmic complexity is important for understanding deterministic chaos, but the condition $K_N/N > 0$ for $N \to \infty$ is just a way of saying that there is no <u>finite</u>-length algorithm that computes the sequence. <u>That is, algorithmically complex numbers are non-computable and therefore algorithmic complexity cannot be used to describe chaos in computation.</u> Algorithmically complex sequences are worse than random; they cannot be defined by any <u>finite</u> number of coin tosses, and there is no way to pass to the limit $N \to \infty$ in any computationally or physically meaningful way.

We have replaced computable maps by automata, using the digitalization indicated in (2). When the map (1) is computable, the decimals $\{\varepsilon_j(n+1)\}$ can be determined from the $\{\varepsilon_i(n)\}$ and $\{\delta_i\}$ by a finite length algorithm, or computer program, for any finite j and any finite n [1,20]. Since computation time is then the fundamental limitation on the prediction of the future from the past, it is interesting to ask whether cellular automata can be used to speed up the computation of chaotic maps.

Above, we have regarded the computable map as the automaton, or processor. Such a processor must handle long binary strings, even for Bernoulli shifts and the logistic map. In the case of the logistic map, the initial condition $x_0 = .001$ is simple, but the dynamics of multiplication of long strings rapidly becomes difficult to handle, either by hand or by machine, and even with quadruple precision you will only be able to perform about six iterations before the machine's floating-point arithmetic begins systematically to introduce errors.

Let us consider an array of relatively simple identical processors where each processor (automaton) is coupled locally to its neighbours. At each "time" n, the jth processor receives symbols or short symbol strings from its neighbours, and updates according to its own internal rule. The rule is the same for all the automata, and since they update simultaneously, this is the idea of parallel processing [24]. The simplest mathematical model of a cellular automaton is perhaps given by

$$\varepsilon_j(n+1) = \varepsilon_{j+1}(n) + \varepsilon_{j-1}(n) \bmod (2) \quad (10)$$

where $\varepsilon_j(n) = 0$ or 1, j is discrete space variable and n a discrete time variable. Although this cellular automaton is too simple, certain two-dimensional cellular automata have been proven capable of "universal computation", while others have been speculated to be capable of universal computation [25]. The idea of universal computation is that such an automaton can in principle compute any computable (i.e. recursive) function. Computable mumbers and functions are exactly what we have been talking about. In order to make direct use of computational universality, it would be necessary to have on hand a "machine language" for a particular universal automaton, that is, an encoding scheme whereby the discretized (digitalized) dynamical system and its initial condition could be written as an initial condition, or "program" for the cellular automaton. The dynamics of the cellular automaton would then compute the orbit of, for example, the logistic map, and to know the map's orbit, you should then know how to decode the cellular automaton. Here, it seems to be a trade-off of complicated dynamics with simple initial conditions for simple dynamics with complicated initial conditions. It is not clear, a priori, that one is more or less complex than the other. Also, is it possible to actually construct a useful machine language for a given universal cellular automaton? If so, how?

Frisch et al. [26] have made a very interesting attempt to model the Navier-Stokes equations via a lattice-gas, or Ising-like cellular automaton. Here, the approximations are largely uncontrolled, so it will be interesting to see whether the model can be used to generate statistics that agree with experiment, or even with more conventioanl simulations.

6. Acknowledgement

I am grateful to T. Riste and Institute for Energy Technology for hospitality at Kjeller, Norway, during spring, 1987, and to Gerd Jarrett for typing this manuscript. I am indebted to J. Palmore, who collaborated with me to produce the ideas discussed above. This work was supported in part by NORDITA.

7 References

* Permanent address, Physics Department, University of Houston, Houston, Texas 77004, USA

1. J.L. McCauley, J.I. Palmore: Phys. Lett. A115, 433 (1986).
2. J. von Neumann: Theory of Self-Reproducing Automata (Univ. of Illinois Pr., Urbana 1966).
3. J.E. Hopcroft, J.D. Ullman: Introduction to Automata Theory, Language, and Computation (Addison-Wesley, Reading, MA 1979).
4. D.E. Knuth: The Art of computer Programming II, Semi-Numerical Algorithms (Addison-Wesley, Reading, MA 1981).
5. R. Shaw: Z. Naturforsch. 36a, 80 (1980).
6. J.I. Palmore, J.L. McCauley: Shadowing by Computable Chaotic Orbits, to be published, Phys. Let. A (1987).
7. H.G. Schuster: Deterministic Chaos (Physik verlag, Mosbach 1984).
8. S. Grossmann, S, Thomas: Z. Naturforsch. 32a, 1353 (1977).
9. C. Beck, G. Roepstorff: Univ. of Aachen preprint (1986)
10. G. Benettin, G.M. Casartelli, L. Galgani, A. Giorgilli, J.M. Strelcyn: Nuovo Cimento B44, 183 (1978).
11. J. Guckenheimer, P. Holmes: Nonlinear Oscillations, Dynamical Systems, and Bifurcations of Vector Fields (Springer Verlag, New York, Berlin, Heidelberg, Tokyo 1983).
12. O.E. Lanford III: in Chaotic Behaviour of Deterministic Systems, ed by Ioos, Helleman, Stora (North Holland, Amsterdam 1983)
13. J.L. McCauley, J.I. Palmore: in Scaling of Disordered Systems, eds. R. Pynn, A. Skjeltorp (Plenum Pr., New York 1985).
14. J.I. Palmore, J.L. McCauley: Statistics of Computable Hyperbolic Systems, preprint (1987).
15. A. Hodges: Alan Turing: The Enigma (Simon and Schuster, New York 1980) pp. 91-110. 16. A.M. Turing: Proc. London Math. Sa (2) 42, 230 (1937).
17. M.L. Minsky: Computation, Finite and Infinite Machines (Prentice-Hall, London 1967).
18. I. Niven: Irrational Numbers, The Carus Mathematical Monographs No. 11 (1956).

19. M. Kac: Statistical Independence in Probability Analysis and Number Theory. The Carus Mathematical Monographs No. 12 (1959).
20. J.L. McCauley: Z. Naturforsh. 42a (1987), in press.
21. S. Waggoner: Math. Intelligencer 7, 65 (1985)
22. P. Martin-Löf: Inf. Control 9, 602 (1966).
23. J. Ford: Physics Today 36, 40 (1983).
24. K. Preston, Jr., M.J.B. Duff: Modern Cellular Automata (Plenum Pr., London 1984).
25. S. Wolfram: Physica 10D, 1 (1984).
26. U. Frisch, B. Hasslacher, Y. Pomeau: Phys. Rev. Lett. 56, 1505 (1986); N. Margolus, T. Toffoli, G. Vichniac: Phys. Rev. Lett. 56, 1694 (1986).

Digital and Analog Approach to Intermittencies and 1/f Noise in a Nonlinear Helmholtz Oscillator

M.A. Rubio, M. de la Torre, J.C. Antoranz, and M.G. Velarde

Departamento de Fisica Fundamental, U.N.E.D.,
Apartado 50.487, E-28080 Madrid, Spain

1. Introduction

In the past few years it has been established by theory and experiment that deterministic chaos can be attained by several routes[1-10]. These include three intermittencies [2,3]. Experimental and numerical evidence exists for intermittencies I and III, and little is really known about type-II [11]. For the type-III intermittency, however, our present understanding is far below the level of sophistication attained in type-I intermittency.

Here, we provide complete digital computation results and analog circuit experimental characterization of type-III and type-I intermittencies in a nonlinear oscillator subjected to external forcing. For the model to be described below we provide the bifurcation diagram, Floquet multipliers, Poincaré and return maps and power spectra (with the predicted low-frequency, 1/f, divergence). In contradistinction with earlier findings [12,13] the predicted 1/f noise [8] appears here unambiguously related to the existence of a purely intermittent single chaotic attractor.

The Helmholtz oscillator refers to the following model problem

$$d^2x/dt^2 + g\, dx/dt + x + x^2 = A \cos(wt) \qquad (1)$$

where t denotes time and x is the unknown, which is related to the ear drum's response to hammer's action and other related phenomena in the ear's interior [14], g is a damping coefficient, A is the amplitude of the forcing term and w is the frequency normalized to the eigenfrequency of the oscillator. Equation (1) corresponds to the motion of a particle in an asymmetric cubic potential, with escape to minus infinity at x=-1.

2. Type-I intermittency

If A is taken as the bifurcation parameter, a typical bifurcation diagram is shown in Fig. 1 for g=0.52 and w=1.2. For $A < 0.5121$, the system oscillates at the frequency imposed by the forcing. Then at A=0.5122, the limit cycle undergoes a subharmonic bifurcation, i.e., a period doubling bifurcation, as can be inferred from the evolution of the Floquet multipliers. A Floquet multiplier crosses the unit circle at minus one. Subsequently at A=0.5152 there is a second bifurcation into an intermittent state depicted in Fig. 2 (a stands for the digital computation and b for the analog experimental result). Thus as we vary the bifurcation (control) parameter there is one and only one period doubling followed by intermittency with no relaminarisation process.

Figs. 3(a) and 3(b) show respectively the first and second iterate of the return map for A=0.517 obtained by stroboscopic method. The second iterate of the return map follows a cubic polynomial near the unstable fixed point. A Forsythe's cubic polynomial fitting yields

$$x_{t+2T} = m_1\, x_t + b\, x_t^3 \qquad (2)$$

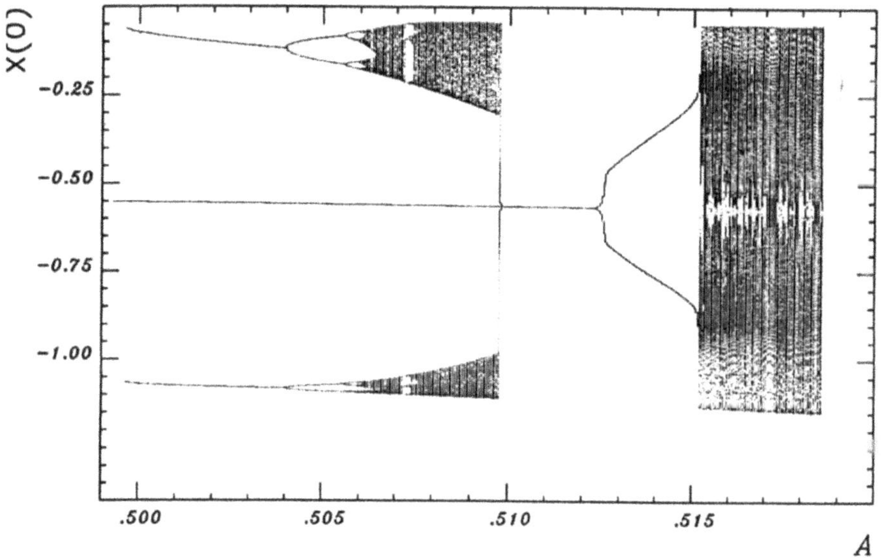

Fig. 1. Bifurcation diagram (using a stroboscopic method) at g=0.52 and w=1.2. The first bifurcation from the primary oscillatory state is a period doubling at A=0.5122, followed by a second bifurcation into an intermittent state at A=0.5152. For A > 0.5195 the system escapes to minus infinity. A different cascade of bifurcations appears surrounding the periodic solution of period T. The behavior of the oscillator is chaotic for some initial conditions for 0.5062 < A < 0.5097. At A=0.5097 the system evolves to the stable T-periodic solution

with b=0.014±0.001 and m_1=1.016±0.001 (for A=0.518), which agrees quite well with the value obtained for the Floquet multiplier (1.015). Taking into account that the last stable periodic solution is 2T, where T is the period of the forcing term, we can classify this intermittency as belonging to type-I.

On the other hand, a complete Feigenbaum cascade can be observed in Fig.1 for a different set of initial conditions. This cascade ends in a different chaotic attractor which coexists with the solution of period T. This strange attractor disappears at A=0.5097, evolving towards the T-periodic solution.

3. Type-III intermittency

By using the same methods as above and changing the parameter values of w and g to w=1.3 and g=0.48, we get the bifurcation diagram depicted in Fig. 4. For A < 0.5609, the system oscillates with period T. At A=0.5610 the oscillator undergoes a subcritical bifurcation driving it to a chaotic state with identical features as shown above.

First and second iterates for the return map are similar to the case of type-I intermittency, but in this case the last stable solution is T-periodic, and as statistical properties are intimately related to the return map, it is clear that the behavior of this system in both cases (type-I and type-III intermittencies) is the same in the chaotic regime. However, in the present case Floquet multiplier crosses the unit circle by minus one, not by one as above. As in Fig. 1, a complete cascade of subharmonic bifurcations is observed, ending in a strange attractor which disappears decaying to the T-periodic solution.

A continuous transition between both scenarios has been observed in parameter space A, g and w.

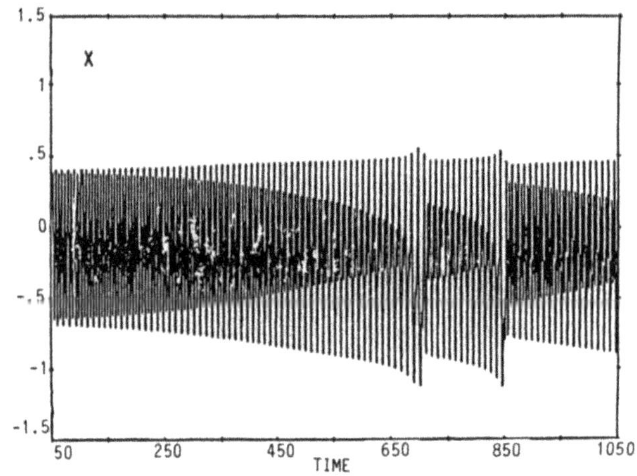

a

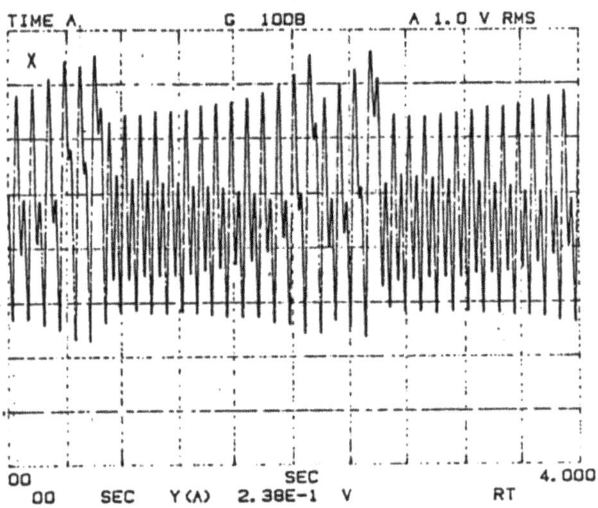

b

Fig. 2. Time evolution of x in the intermittent regime: a) digital computation for A=0.518, g=0.52 and w=1.2, b) analog simulation for A=0.5, g=0.52 and w=1.3

4. Low-frequency behavior

Figures 5(a) and 5(b) provide another important feature to point out, i.e., the low-frequency, 1/f, divergence in the power spectrum. Figure 5(a) corresponds to the digital computation whereas Fig. 5(b) is the experimental evidence obtained with the analog computer. Due to the narrowness of the chaotic behavior and the intrinsic noise of the electronics it is quite difficult to stabilise the analog circuit for a time long enough to achieve averaging of the power spectra in order to show the low-frequency noise generated by the dynamics of the system.

For the set of values g=0.52, w=1.31 and A=0.501, corresponding to a bifurcation diagram similar to the one displayed in Fig. 4, we obtain the spectrum shown in Fig. 5(b). The low-frequency noise in the periodic state is three orders of magnitude lower than in the intermittent regime, thus showing that the 1/f divergence is indeed genuine of the dynamics of the intermittency of the system, which is a beautiful confirmation of the available theoretical predictions [2,8] and numerical calculations [15,16].

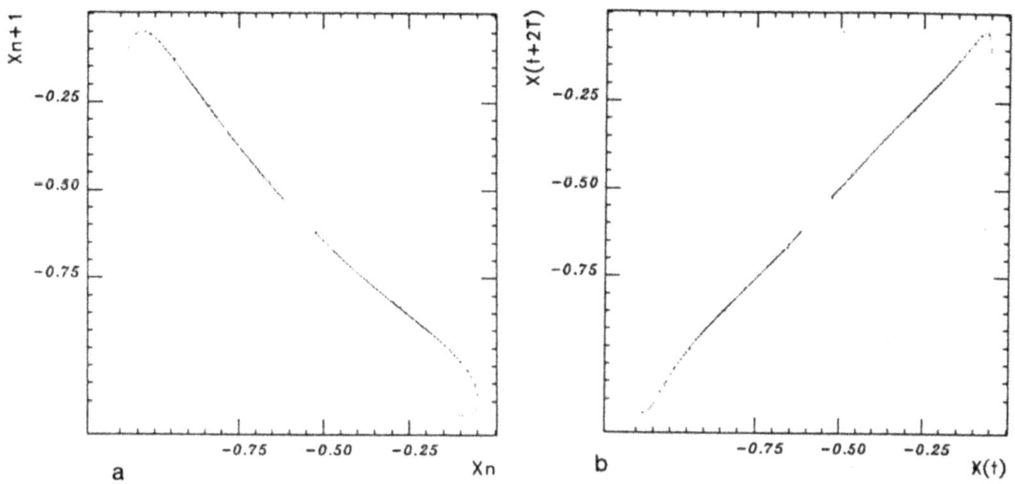

Fig. 3. Return maps of x for A=0.517, g=0.52 and w=1.2. a) First iteration of the return map. The slope at the intersection with the bisector line is minus one, b) second iteration, the slope is now slightly higher than plus one

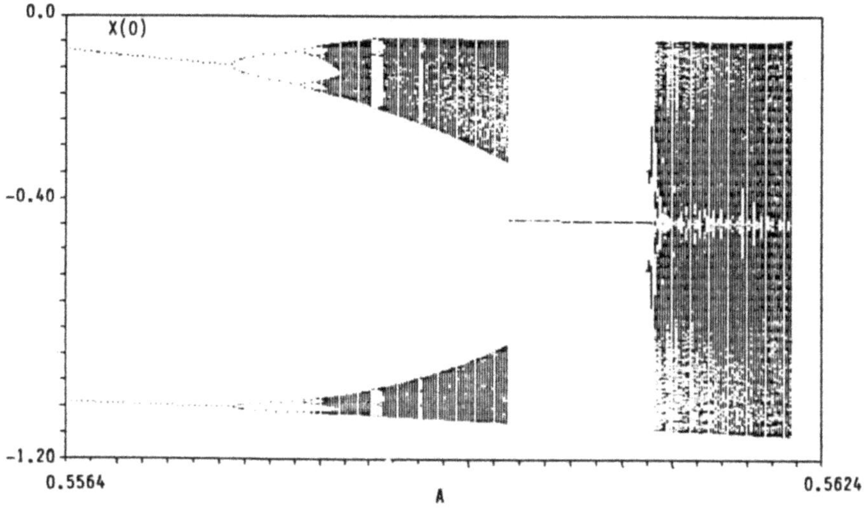

Fig. 4. Bifurcation diagram at g=0.48 and w=1.3. The only bifurcation from the T-periodic state is an intermittent chaotic state at A=0.5610

Acknowledgements

The authors acknowledge fruitful discussions with Profs. H. Haken, E. Holzhauer, J. Tredicce and Dr. H. Bunz and access to unpublished research results. They would also like to acknowledge hospitality at I.N.O. (Florence, Italy) (M.A. Rubio) under a NATO Grant, and at Los Alamos Nat. Lab. (J.C. Antoranz and M.G. Velarde). This research has been sponsored by the Stiftung Volkswagenwerk.

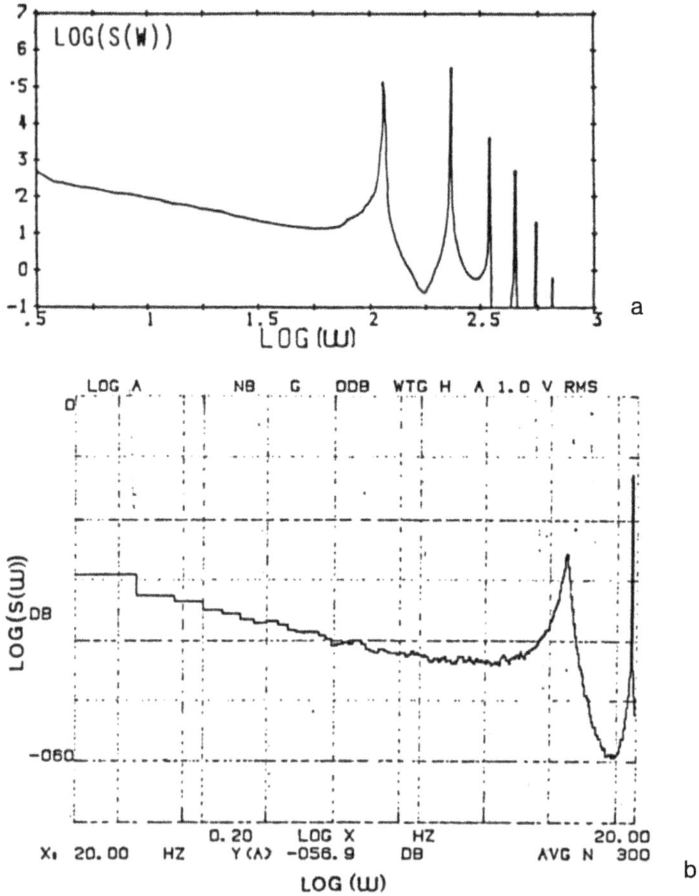

Fig. 5 Power spectrum of x, a) digital computation for A=0.518, g=0.52 and w=1.2. The slope at the low-frequency region is -1.2, b) analog simulation for A=0.5, g=0.52 and w=1.3. The slope in this case is -1.1.

References

1. H. Haken, Advanced Synergetics, Springer-Verlag, Berlin, 1983.
2. P. Bergé, Y. Pomeau and Ch. Vidal, L'ordre dans le chaos, Hermann, Paris, 1984.
3. Y. Pomeau and P. Manneville, Commun. Math. Phys., 77(1980), 189.
4. J.E. Hirsch, B.H. Huberman and P.J. Scalapino, Phys. Rev. A26(1982), 2117.
5. Th. Geisel and J. Nierwertberg, Phys. Rev. Lett. 48(1982), 7.
6. H. Bunz, H. Ohno and H. Haken, Z. Phys. 56B(1984), 345.
7. M. Dubois, M.A. Rubio and P. Bergé, Phys. Rev. Lett., 51(1983),1146.
8. A. Ben-Mizrachi, I. Procaccia, N. Rosenberg, A. Schmidt and H.G. Schuster, Phys. Rev., A31(1985), 1830.
9. M.G. Velarde and J.C. Antoranz, in Instabilities and dynamics of lasers, R.M. Boyd, L.M. Narduci and M.G. Rayner editors, Cambridge University Press, Cambridge, 1985.
10. E. Holzhauer and E. Rauchle, private communication and report at an International Synergetics Conference (unpublished), Elmau, May 1984.

11. P. Richetti, F. Argoul and A. Arneodo, Phys. Rev, A34(1986),726.
12. F.T. Arecchi and F. Lisi, Phys. Rev. Lett., 49, (1982), 94, and F.T. Arecchi and A. Califano, Phys. Lett., 101A(1984), 443.
13. R.F. Miracky, M.H. Devoret and J. Clarke, Phys. Rev., A31(1985), 2509.
14. H. Helmholtz, Die Lehre von der Tonempfindungen als Physiologische Grundlage fur die Theorie der Musik, Friedrich Viewveg, Braunschweig, 1870.
15. Y.H. Kao, J.C. Huang and Y.S. Gou, Phys. Rev., A34(1986), 1628.
16. E.G. Gwinn and R.M. Westervelt, Phys. Rev., A33(1986), 4143.

Part VII

Robotics

Towards Automatic Error Correction in Robots: Inferring the Task from the Program

V. Caglioti and M. Somalvico

MP-AI Project, Milan Polytechnic Artificial Intelligence Project,
Dipartimento di Elettronica, Politecnico di Milano,
Via Ponzio 34/5, I-Milan, Italy

1. INTRODUCTION

During the early 1970s, specific programming languages for robots were developed in order to improve their versatility. More recently, many different artificial intelligence techniques have been applied to robots, improving their functional flexibility.

The occurrence of errors is usual in industrial applications of robotics. The problem deriving from it can be depicted as follows: An (assembly) task has to be accomplished by a robot. A human programmer constructs an algorithm and assigns it to the robot in order to attain the desired world modifications, according to the task. In order to expedite both the program writing and the program execution, many improbable and undesired world evolutions are not taken into account in the program.

The robot, during the execution of the assigned program, can be interrupted within an assembly process by an error caused, e.g., by a defective object, by a wrongly positioned object, by the absence of an object or by an undesired collision between objects. Therefore, when such an unpredicted event happens, the correct execution of the program may be jeopardized or even stopped. Currently, at this point, a human rearrangement of the objects in the robot's working area is necessary to allow the continuation of the program execution.

The ultimate goal of the research on automatic error correction is to reduce the need of human presence near the robot during the execution of an assigned plan, whatever the assigned program is.

A way to pursue this purpose is to provide the robot with the ability to self-correct some execution errors. Typical behaviour of the robot provided with this new ability would then be the following. During the execution of an assigned program an error occurs. The robot detects it. The robot sensors are then used to acquire information about the actual present (error) state. Then a strategy is planned in order to correct the error that has occurred. Finally this strategy is executed and, if well accomplished, the normal execution of the assigned program can be resumed.

One more activity must take place during the off-line phase, as a prelude to the above on-line correcting activity: the identification of the goal states to be reached by the correcting strategy. These goal states must be determined on the basis of both the particular task to be accomplished by the robot and the particular error state in which the execution has been interrupted.

The main source of information about the task consists of the assigned program. In general, robot programming languages allow instructions that are expressed in terms of the movements of the manipulator rather than in terms of the world modifications produced by these movements.

A three-module architecture for a correcting system has been proposed in [1], including a task-inference module, an error detecting module and a correcting strategy planning module. In [1] and [2] it was assumed that some high-level descriptions of the task were available (like e.g., "Screw S1 with screw S2 and block B1 at position P1!"). In [3] the robot action repertoire has been limited. A monitoring system has been proposed [4], in which some monitoring actions are interposed among those of the originally assigned program. Therefore an "augmented program" is executed. Among the monitoring actions proposed in [4] there are control sensory detections and statement constructions that create facts providing information about how the planned execution is actually proceeding.

In almost all the mentioned approaches, a quite concise description of the evolving world state is constructed, so that the main correcting strategies are based on repetitions of the attempts to execute some of the actions of the assigned program. This makes it difficult to cope with programs containing branches or cycles.

The present contribution illustrates a knowledge-based approach to the inference of the task from the program, which allows the identification of suitable goal states for the correcting strategy. The task inference module analyzes the assigned program, constructing a general description of the evolving world state. This description contains statements about positions, contacts and constraints among object. This approach can deal with a quite generic program, provided it belongs to the assembly problem domain, whose structure is not necessarily linear but may show branches and loops.

Topics such as the interpretation of sensory data [7], their translation from a signal form into symbolic information [6], automatic reasoning about physical systems, and automatic robot planning [5] are relevant for the research on automatic error correction.

The availability of a correcting system based on the task inference mechanism illustrated below would be convenient for two reasons:

- A certain amount of human programming work is saved by avoiding having to tackle some problems like, e.g., defective objects, wrongly positioned objects, etc. These problems would be left to the correcting system, which would deal with them a posteriori, i.e. only once they had arisen.

- If the correcting system is able to work with a generic task (though belonging to a particular problem domain), the programmer's work saving can be multiplied by the number of programs that can be written that are covered by the domain.

The contribution is organized as follows : in Sect. 2 the relationships between the problem of automatic error correction and task inference are shown, in Sect. 3 a task representation formalism is proposed, in Sect. 4 the task inference process is illustrated, in Sect. 5 the goal states identification mechanism is depicted and in Sect. 6 some conclusions are outlined.

2. THE PROBLEM OF AUTOMATIC ERROR CORRECTION: RELATIONSHIP WITH TASK INFERENCE

Errors can occur during the execution of an assigned program due to both the finite precision and the limited definition of the world knowledge involved in the program generation. Therefore not all the possible world evolutions are foreseen by the generated program. The lack of both precision and definition in the world knowledge involved in the program generation is equivalent to a "hypothesis assumption". These hypotheses are often implicitly assumed by the programmer, because he relies on human supervision of the robot activity.

By means of this hypothesis assumption, the programmer can avoid writing many long control and recovery routines. Instances of such hypotheses are the following :

(objects are not defective)
(declared object characteristics are true)
(assumed object positions (and orientations) are true)
(constraint between connected objects are in accordance with the geometrical characteristics of the objects).

A program execution is "correct" if the assumed hypotheses are true during this execution. We shall refer to an evolution that derives from a correct execution of the program as a "desired" evolution.

Errors represent deviations of the actual evolution from the set of desired evolutions. Before the correction of an error that has occurred can be attempted, three preliminary activities have to be performed :

1. Some of the knowledge that had been neglected in the hypothesis assumption is retrieved, e.g., rules like :

```
IF ((defective -object) )
THEN
   ((proposed substitution -object) ).
```

2. Some knowledge about the actual (error) state is extracted, e.g., statements like :

```
(defective Block1)
(class Block1 peg)
(class Block2 holed block)
(state that would have been reached state1).
```

3. Some knowledge about the particular task to be accomplished correctly is employed, e.g., rules like :

```
IF ((class -Block1 peg)
    (class -Block2 holed block) )
THEN
   ((desired constraint -Block1 -Block2 -direction1 -state1)
    (desired constraint -Block1 -Block2 -direction2 -state1))

                        or

IF ((class -Block1 cuboid)
    (class -Block3 plane) )
THEN
   ((desired constraint -Block1 -Block3 -direction3 -state1)).
```

A knowledge of the actual state directly provides a description of the starting state of the correcting strategy planning problem. The task knowledge allows the determination of goal states for the correcting strategy planning problem, e.g., the statements :

```
(desired constraint Block1 Block2 direction1 state1)
(desired constraint block1 Block2 direction2 state1) .
```

The retrieved knowledge provides some proposals of ways of solving the planning problem, e.g., in the case of a defective object, its substitution is proposed in the hope of reaching one of the possible goal states.

Once an error has been detected during the execution of the assigned program, the constructed descriptions of both the starting state and the goal state(s) of the correcting strategy allow the planning problem to be well posed. The retrieved knowledge avoids the infinite reiteration of the same action that led to the error. That is: the retrieved knowledge allows the planning problem to be solvable in practice.

The task knowledge can be processed off-line, by means of an analysis of the assigned program, together with some knowledge about both the set of the possible initial states of the execution and the relevant physical and geometrical characteristics of the objects that will be manipulated by the robot.

Therefore, a correcting system can be composed of three modules :

1. An off-line operating module that analyzes the assigned program together with other necessary information infers the task to be accomplished correctly, and translates it into a suitable form that allows the identification of goal states for subsequent planning of correcting strategies.

2. An on-line module that directs the whole on-line activity. It acquires knowledge about the actual error state (and possible about the error cause), and decides when to give the control to the next module, which attempts the execution of a correction strategy.

3. An on-line module that selects some goal states (in accordance with the task representation inferred by the first module) and plans their attainment possibly taking into account the knowledge acquired by the second module.

The present approach deals specifically with the task inference activity and with the goal identification method.

By means of the task inference activity, a task representation is extracted that corresponds to the analyzed program. The association of a task representation with a program is done in accordance to some "criteria". These criteria descend from the chosen task representation formalism, which will be described and justified in the next section.

In the present approach, attention is paid specifically to the following problems :

a. Once an error has occurred, the current desired evolution is abandoned definitively. The error state does not belong to any of the desired evolutions. Some other evolutions are intuitively considered as "acceptable", from the programmer's point of view. For a useful task representation it is not sufficient to determine just the set of desired world evolutions; the acceptable evolutions have also to be taken into account. Therefore the task representation formalism is to be chosen so that the set of acceptable evolutions can be uniquely determined.

b. The task representation is used in the determination of goal states to be reached by any correcting strategy. The assigned program, which constitutes the main information source to be employed while inferring the task, is expressed in terms of "actions" rather that in terms of world modifications. Therefore, there must be a phase of the task inference during which robot actions are translated into world modifications. The physical state of the world is described in terms of statements about positions, contacts and constraints among objects (for the assembly problem domain).

c. In the general case, the program to be analyzed for the task inference can contain sensory actions. By one of these actions, the program associates the measured value of a sensed physical quantity (relating to the current world state) with a memory variable. Subsequent program actions can prescribe tests on the measured quantities. Different truth values of test predicates can lead to different branches of the program. The program structure can also show cycles, which originate from tests, too. In order to account for these cases, the task representation allows multiple states for each program step (each step corresponding to the execution of a program action).

3. THE TASK REPRESENTATION FORMALISM

In the following, the term "evolution" will recur. An evolution deriving from the execution of a sequence of actions is the discrete trajectory of the world state. Each step at which the state is considered corresponds to the end of the execution of an action of the sequence. As mentioned in problem c of the last section, at each step of a program execution there can be many different world states. Therefore, a whole set of different evolutions can derive from the execution of a (single) program.

In accordance with the discussion of problem a of the last section, let us consider a desired evolution (deriving from the execution of the assigned program). For each desired evolution, there is a set of "acceptable" evolutions (from the programmer's point of view). For instance, an evolution composed of the combination of :

1. an initial part of the considered desired evolution interrupted by an error;

2. a partial evolution, deriving from the successful execution of a correcting strategy;

3. a continuation of the evolution, deriving from the resumption of the program execution;

is an "acceptable" evolution, though not a desired one (due to the error occurrence).

The task representation formalism has to be chosen in order to include all (and only) those evolution characteristics that are relevant from the point of view of the programmer. The particular representation must include conditions on the relevant evolution characeristics that are sufficient to decide whether a generic evolution is acceptable or not, from the programmer's point of view.

Now a set of relevant evolution characteristics is proposed, and the consequent task representation formalism is derived.

One can use the term world evolution to refer to the evolution of the following system : the system composed of the robot and that part of the world that is accessible to the robot by means of its actuators or its sensors. We suppose that this system can interact with the external world only through communication channels. Among these communication channels there are :

- those behaving as inputs, like boxes and tapes carrying objects from the external world towards the system;

- those behaving as outputs, like boxes and tapes carrying objects from the system to the external world.

Robot actions are not considered as inputs for the system, because they are applied by the robot to the (rest of the) system. It is assumed that the system evolution cannot derive from a cause that is uncorrelated with both the robot and the system input. In this sense, the User belongs to the external world, and is assumed to have access to the system only through its communication channels; practically, he can perceive and evaluate an evolution of the system only by evaluating its input and output characteristics. The internal evolution of the system is "invisible" to the User.

An extension of the meanings of input and output can be made while dealing with our system. According to basic systems theory concepts

- input can be used to refer to an entity that affects the system evolution without being affected by it;
- output can be used to refer to an entity that is affected by the system evolution without affecting it.

Therefore, the initial state of the evolution can be considered as an input of the system, while the final state of the evolution can be considered as an output of the system.

Input interactions between the system and the external world occur when new objects enter the system. Output interactions occur when objects leave the system. Input interactions are described in terms of class, physical and geometrical characteristics, positions, contacts, and constraints of objects that enter the system. Output interactions are described in a similar way.

The extended input and output characteristics are part of the proposed relevant characteristics of a generic evolution, that is :

- the class, geometrical and physical characteristics, positions, contacts and constraints of the objects that either are present in the system at the initial state or enter the system via input;

- the class, geometrical and physical characteristics, positions, contacts and constraints of the objects that either are present in the system at the final state or leave the system via output.

Since the system is evolving along a discrete time scale, another evolution characteristic that is perceivable by the User is the chronological order of the interactions of the system.

The above characteristics, together with the chronological order of the system interactions, are referred to as the "input and output configuration" of the evolution. This input and output configuration constitutes the relevant evolution characteristics that will be evaluated, once the task representation of a given program has been determined, in order to decide whether the considered evolution is acceptable or not. Of course, there are many evolutions that have the same input and output configuration.

The relevant evolution features have been proposed, now a description of the consequent task representation formalism is given.

Given an assigned program, the task representation formalism must allow the decision whether a generic evolution is acceptable, or not, by means of the evaluation of the relevant characteristics (i.e., the input and output configuration) of this evolution. To this end, the desired evolution, deriving from correct executions of the assigned program, are of course classified as acceptable.

Starting with the set of desired system evolutions, deriving from correct executions of the assigned program, the task representation is constructed by listing the input and output configurations of all the desired evolutions. A generic evolution is acceptable if its input and output configuration belongs to the set of the i/o configurations of the desired evolutions.

In analogy with the many different allowed states at each step of the program execution (see the beginning of this section), in a generic program many different input configurations are possible (e.g., due to variable characteristics of an object that enters the system at a certain step).

The task representation formalism has been illustrated. The method is now described by means of which the task representation is determined, starting from the assigned program. This method will be illustrated for the case of absence of output interactions, i.e., objects are not removed from the system until the final action has been executed.

4. THE CONSTRUCTION OF THE TASK REPRESENTATION OF THE ASSIGNED PROGRAM

The set of the i/o configurations of the desired evolutions has to be constructed, starting from the assigned program. Further information is needed, like the description of the interactions (i.e., the objects that enter the system after the initial state) and the description of the objects that are already present in the system in the initial state.

The i/o configuration of the desired evolutions are determined in parallel with the construction of the desired evolutions. These are constructed by simulating the "correct" execution of the assigned program.

In order to attain only the "desired" evolutions from the simulation, rather than all the "possible" ones, the hypotheses that were assumed by the programmer are embedded in the world knowledge employed in this phase. Furthermore, in order to expedite the simulation, some rules are used that may play the role of restricting rules for the programmer. But these rules are effective for the simulation. Instances of these rules are :

- movements do not stop before their arrival positions have been reached.
- during a movement action, an object may only either remain at rest or move rigidly connected with the moving reference frame attached to the robot hand.

The simulation of the correct execution of the program leads to the construction of the set of desired evolutions. It is carried on by reiterating a step.

During the generic step in which the execution of an action is simulated, all the arrival states are determined by analyzing the transitions deriving from the application of the action to each of the starting states. The next step uses these arrival states as starting states for the simulation of the subsequent action.

At the end of the simulation step, each arrival state inherits from its corresponding starting states :

- their connection(s) with initial state(s);
- their connection(s) with partial input configuration(s); (partial input configurations are represented by statements, relating to only those input interactions that precede the currently simulated action).

If the simulated action is an input interaction, each arrival state adds, to the inherited partial input configurations the "local" input configuration (relating just to the current input interaction) associated with the considered transition.

Therefore, each state is connected with both the initial states and the partial input configurations that are compatible with it. At the end of the simulation, the set of the desired i/o configurations is constructed by joining up each final state with all the input-configurations and initial-configurations that are compatible with it. This set constitutes the task representation of the simulated program.

To allow computation, the system state is represented in a well-defined way. The state is represented by a set of statements, describing the positions of objects, their contacts and their constraints. For instance :

(position Block1 position1 state1)
(position Block2 position2 state1)
(contact Block1 Block2 contact position state1)
(constraint Block1 Block2 impeded relative motion direction state1).

Other statements describe object characteristics, connecting them with object identity. For instance :

(class Block1 holed block)
(friction coefficient Block1 Block3 0.2)
(diameter block1 5).

Other statements describe memory variables. They do not directly concern the physical state of the world, but they are effective for simulation :

(memory var x number 3.4)
(memory var y logical t).

The simulation is performed by a rule-based software system. During each simulation step, given both the action to be simulated and the starting state, rules are used to determine the desired arrival states. The rule firing results in both the deletion of some statements (relative to the starting state) and the creation of new statements (relative to the arrival state). Rules express geometrical and mechanical knowledge, in addition to the programmer's hypotheses and to the programming restricting rules.

The connections of each state, with its compatible input-configurations (and with its compatible initial states), though not useful for the task representation in the case of no final state, are employed during the goal identification for the correcting strategy. The goal identification is illustrated in the next section.

5. THE GOAL IDENTIFICATION MECHANISM

Once an error has occurred, the current desired evolution is interrupted. The complete evolution resulting from the connection of the correcting strategy to the interrupted part of the evolution must be acceptable, i.e., its i/o configuration must belong to the set of the desired i/o configurations.

The initial state of the complete evolution is the initial state of the interrupted evolution. In addition, a partial input configuration has been constructed during the interrupted part of the evolution.

The complete input configuration, resulting from the connection of the configurations of the subsequent input interactions to the partial input configurations of the interrupted evolution, must be an allowed one. Furthermore, the final state of the complete evolution must be compatible with both the complete input configuration and the initial state.

It is understood that if an object substitution is needed, the partial input configuration is modified by changing part of the partial configuration, to be precise, that part concerning the input interaction by which the substituted object entered the system.

The following goal indentification method derives from the above requirements. We suppose that the error occurred just before the step S_0 was reached. A subset of S_0 would have been the set of the predicted arrival states at the step.

A first branch of the correcting strategy is constructed by planning the attainment of the previous subset of goal states. We call it s_0. If an input action is prescribed in this branch, with allowed results that are not compatible with any of the goal states of s_0, then another branch has to be generated, starting from this input action.

The goal states of the new branch are chosen as follows. The planner looks for a step S_1 that satisfies two requirements :

1. S_1 is subsequent to S_0 in the assigned program.

2. S_1 contains at least one state s_1 that is compatible with both the initial state and the new partial input configuration.

This branch generation will be iterated until all the possible results of input interactions that are prescribed in the current branch are compatible with one of the current goal states.

The goal states of the different branches of the correcting strategy are determined. This method requires that the given program is "structured". Therefore no unconditioned jumps are allowed. Branching is only possible with "IF <condition> THEN <action1> ELSE <action2> " -like instructions, while cycles are only possible with "WHILE <condition> DO <action>"-like instructions.

6. CONCLUSIONS

The illustrated paradigm allows task inference for a wide class of assembly programs. Non-linearly structured programs containing sensorial tests leading to branches and cycles are analyzed. The set of the possible initial states and the (possible variable) characteristics of objects entering the system constitute additional required information.

The task inferring module has been realized. It is composed by some programs written in Franz-Lisp and by a rule-based program written in YAPS. This module analyzes robot programs written in VML. Due to experimental requirements, rough software modules are being realized both for the directing module of the correcting system and for the correcting strategy planning module.

The chosen state representation has proven effective even for simulating concurrent assembly programs. In these programs, many processes run in parallel.

The next step will be the study of suitable goal identification mechanisms for error correction in concurrent programs, in order to allow error correction in parallel with the continuation of the program execution.

This aspect would be economically relevant in industrial robot work stations: let us suppose we have a robot work station placed at an intermediate point of a global production line. When an error occurs in the robotic cell, the work station activity must be interrupted in order to allow error recovery. In the meantime, the part of the production line that lies below the robot work station cannot be stopped. Therefore, many units of the partially assembled product have to be always ready to be utilized. Near the robot work station, dozens of these partially assembled parts are stored, waiting for an error occurrence. They constitute a considerable amount of unproductive capital.

This effect can be considerable reduced if error correction can be carried on in parallel with the program execution.

REFERENCES

[1] G. Gini, M. Gini, M. Somalvico, "Emergency Recovery in Intelligent Robots", Proceedings of the 5th International Symposium on Industrial Robotics (1975)

[2] G. Gini, M. Gini, M. Somalvico, "Program Abstraction and Error Correction in Intelligent Robots", Proceedings of the 10th International Symposium on Industrial Robotics (1980)

[3] L. Friedman, "Robot Learning and Error Correction", Proceedings of the 5th International Joint Conference on Artificial Intelligence, (1977)

[4] R.E. Smith and M. Gini, "Robot Tracking and Control Issues In An Intelligent Error Recovery System", Proceedings of the 1986 IEEE International Conference on Robotics and Automation, S. Francisco (1986)

[5] E. Mazer, "Geometric Programming of Assembly Robots", Proceedings of the International Meeting of Advanced Software in Robotics, Liege (1983)

[6] T.O. Binford, "Survey of Model-Based Image Analysis Systems", International Journal of Robotics Research, vol. 1, n.1 (1982)

[7] R. Cassinis, "Sensing System in Supersigma Robot", Proc. IX ISIR, Washington (1979)

Index of Contributors

www.ingramcontent.com/pod-product-compliance
Lightning Source LLC
Chambersburg PA
CBHW061208240926
51846CB00006B/1152
9783642730917